Beyond Semiotics

Beyond Semiotics
Text, Culture and Technology

NIALL LUCY

CONTINUUM
London and New York

Continuum
The Tower Building, 11 York Road, London SE1 7NX
370 Lexington Avenue, New York, NY 10017-6503

First published 2001

© Niall Lucy 2001

British Library Cataloguing-in-Publication Data
A catalogue record for this book is available from the British Library

ISBN 0-8264-4932-8 (hardback)
    0-8264-4933-6 (paperback)

Library of Congress Cataloging-in-Publication Data
Lucy, Niall.
        Beyond semiotics—text, culture and technology/Niall Lucy.
            p. cm.
        Includes bibliographical references and index.
        ISBN 0-8264-4932-8—ISBN 0-8264-4933-6 (pbk.)
        1. Semiotics. 2. Culture. 3. Communication and technology. I. Title.
    P99.L83 2001
        302.2—dc21   00-065827

Typeset by Paston PrePress Ltd, Beccles, Suffolk

Printed and bound in Great Britain by Martins the Printers Ltd,
Berwick upon Tweed

Contents

Acknowledgements

As always, I am grateful to friends and colleagues for taking an interest in my work and commenting on the manuscript. This time around I want to thank Robert Briggs, Martin P. Casey, Lisa Gye, Steve Mickler, Deborah Robertson, Horst Ruthrof, Darren Tofts and McKenzie Wark. I am also deeply grateful to Janet Joyce at Continuum, whose patience and support saved the day; to Gillian O'Shaughnessy, whose friendship has saved more days (and more besides) than I could hope to return; and I thank Murdoch University for granting me some time away from teaching in which to write the book.

Several chapters are based on previously published material, and I thank the editors of the journals in which those earlier versions appeared: Chapter 3 in *Senses of Cinema* (2000), Chapter 4 in *Social Semiotics* (1994), Chapter 5 in *Social Semiotics* (1992), Chapter 6 in *Essays in Sound* (1995) and Chapter 9 in *Semiotica* (1993).

For the world's best sister, Judith Lucy

Introduction: Chance Encounters

I wonder what the chances were (they must have been astronomical) against it being you, wherever you are, you of all people, who should happen to be reading this right now. Try as I might to nullify the mystery of this chance event (*I am but a publishing function, you are just a sale*), still I can't get over the absolute improbability that at this very moment it is you, out of everyone, who is reading me. At *this* very moment, *you*. Needless to say, you arouse my curiosity, though of course the chances are that we will never meet and nothing else will pass between us. Believe me, there's something more than a little overwhelming about that, something all too much about the fact that beyond this encounter (outside the text, as it were), you and I seem destined to remain worlds apart, all but oblivious to one another except for an obscurely felt sense of wanting, perhaps, to look into each other's eyes.

For if not *you* and *I* (to retain for now the existential force of those italics), and both of us together here and now, what are the alternatives? Must we be forced to deny that there is anything mysterious whatsoever about me having written a book that got published which you happen to be reading at the moment? Business as usual, in other words. In market terms, 'you' fit a demographic to which 'I' am writing – who cares about the colour of our eyes?

Either, then, the sheer happenstance of our coming together right here and now, or the absolute mundanity of that event? Are these the only choices?

If so, it would seem a decision remains to be made between the fact of our singularities (the irreducible singularity of *my* identity and of *yours*) and the generality of the event in which we find ourselves conjoined, a generality whose structure would deny 'us' any identity beyond the purely functional (thus reducing 'you' and 'I' to available subject positions able to be occupied by any number of singularities, including you and I). Or rather, in those terms, nothing remains to be decided at all since it is obvious that you and I are incidental to the general structure of an event that may be put as *someone wrote a book that someone else is reading*. Perhaps I, as the book's author, retain a certain indispensability in comparison to you, but all the same it would seem insignificant that it's *you* and *I* whose coming together happens to constitute a so-called event at this very moment. Moreover, by the way, which very moment is that? For surely my present is your past; even for me indeed the present is turned into the past every time I come back to this text and read over what I wrote the night before, editing it and adding to it, night after night. And who knows . . . by the time this book is 'finished', the most contemporary fragment of it could be the first sentence rather than the last?

We begin, then, in the middle of things, with our sense of feeling individual, while also knowing that our individuality is inscribed in a set of relations with other individuals within a community or culture comprising various institutions (such as a publishing industry), conventions, beliefs and so forth, all qualifying what it means to be 'us' in terms of any absolute singularity. The talk today is that those relations have been turned on their head by 'technology', such that contemporary media and communications technologies are responsible for producing a 'false' culture of 'alienated' individuals due to the 'distancing' effects of 'impersonal' forms of 'postmodern' contact. A child sitting at a home computer in Idaho can strike up a 'conversation' with another child in some other part of the world (let's say an African village, to give the utopic case for technology its strongest due) via an internet connection, but does he or she know even the names of the kids next door? The more we're technologically connected (so the dystopic case against technology goes), the more we are actually 'disconnected' from who we are and one another, disconnected from some sense of 'real' being that derives only from within 'authentic' culture.

History is full of doom-saying about the parlous state of cultural affairs at just about any given moment; it is also full of ecstatic pronouncements on the cultural benefits to be had from new technologies. In this book I'm a good deal less concerned with over-excitement at the prospect of a bright new future than with po-faced finger-pointing at who's responsible for having lost our better past. More often than not, the finger points to postmodernism. It's not a surprise any more that 'pomo' is the default-set enemy of the guardians of culture; what is surprising is that those guardians should include advocates of semi-

otics. The Canadian semiotician Marcel Danesi, for example, understands this to be 'a time when a belief in the *soul*, which in the "age of the
gods" was motivated by a universal instinctive religious feeling, has
been replaced by materialistic and mechanistic notions' of self-interest
expressing a general 'disenchantment with the traditional narratives of
our culture'.[1] Again, in a book on 'postmodern' semiotics, M. Gottdiener
defines ours as a *lapsed* age of 'image-driven culture' in which 'the
grounds of meaning in daily life' are just too unsettling for his comfort:

> Mass cultural production diverts our ability to place deep-level
> understandings on the images that are constantly valorized for us
> by that industry [the culture industry]. In particular, there remains
> an often implicit acknowledgement that postmodernism, more so
> than previous cultural modes, subverts the desire for authenticity
> in everyday life. It is, in fact, the enemy of authenticity, posed as it is
> to pounce on any cultural form and strip its deep meanings bare for
> the benefit of handling a decontextualized and easily manipulable
> signifier as some new image or façade.[2]

It's a worry that this sort of stuff, written in the name of semiotics,
abounds, because it's bound to give semiotics a bad name. Rest assured,
however, we won't be lamenting lost authenticity or warning against
cultural disenchantment here. I prefer to think that once you've done
'paradise lost' and 'paradise regained', there is still much left to be said
about culture. Indeed it's because I don't think culture was given to us
by God (or the gods) that I don't think we can imagine it as having been
paradisal then and pretty God awful now! This is also why I do not
think technology should be either hysterically hyped up or scornfully
heaped upon.

Now in so far as Danesi's and Gottdiener's take on contemporary
culture owes anything to semiotics, we need to go 'beyond' semiotics in
order to look at culture differently. I elaborate on this in Chapter 2, but
the point is clear enough already: one of the meanings of the title of this
book is that 'beyond semiotics' refers to ways of approaching contemporary culture (which is intricately caught up, I argue, in questions
concerning text and technology) without conceiving it as something
'lost' that has to be 'regained'. While semiotics cannot be held solely
responsible for that conception, neither can the position put by Danesi
and Gottdiener (or that of Marshall Blonsky, as discussed in Chapter 2)
be seen as a complete departure from Saussure's promised science of
culture based on his radical concept of the conventional or arbitrary
structure of the sign.

This is not to assume that you're familiar with semiotics either generally or in detail, or even at all. Indeed throughout the book I have added a gloss or note (wherever it seemed confusion could otherwise reign) on key terms and concepts relating not only to semiotics but also, among other subjects, deconstruction. From the title it is clear that *Beyond Semiotics* is not an introduction to semiotics; yet it's not an attack on it either. While I want to go beyond the miserabilist line on culture which is sometimes taken in the name of semiotics (though of course not only in that name), I neither want nor think it possible to go beyond the Saussurean concept of the sign in the sense of being able to replace it or, these days, even think without it. This touches on a more difficult sense of 'beyond semiotics' in which the movement away from semiotics is at the same time internal to semiotics itself. This is to say that Saussure's concept of the sign opens a space for deconstruction, which is one way 'beyond' semiotics (the way taken here) but not a way to get 'outside' it since in fact it comes from within, belonging to semiotics if only as a potential force or direction. I discuss this at greater length in Chapter 5, but here let me say that to go beyond semiotics while remaining within it is to open (or re-open) Saussure's sign to indefinite, disclosural, nonoppositional effects or possibilities which are internal to it, in potential. As a science (or simply as a discipline), semiotics has by and large refused the sign's internal forces of disruption, preferring to set limits on the meaning of any sign according to a set of systematic principles of interpretation. Such limits (like any limits) mark a division, in this case between what is inside and what is outside any sign. Hence to move beyond semiotics is to ask the question – what grounds the structure of any inside/outside opposition? Or, what are the grounds (the limits) of limits? This is not a question for semiotics, but neither is it given much attention generally, 'outside' semiotics as it were. Hence we may say that semiotics itself belongs within some larger structure that might be called a way of thinking which is characterized by certain assumptions concerning the limits of any difference or the grounds of any opposition and thus the nature of identity. It is not only semioticians who hold to a certain certitude when it comes to knowing the identity of a sign, a structure or a difference; we all do. Each of us knows his or her own identity; knows that there is a structure to television soap operas and that it differs from the structure of quiz shows; knows that the sign 'Nike' is different from the sign 'Adidas'. Nothing could be more certain.

In a sense, that's true. But the problem is that certitude runs the risk of being certain about the wrong things. Racial superiority, the cultural validity of female circumcision, the economic legitimacy of unemployment, the endorsement of state-funded vocational studies at university – none of these positions is held in any doubt by those who hold them. How to shift them, then? Simply by opposing one certitude to another, by replacing this certitude with that? Well yes, it is certainly necessary sometimes to act in such a way; and it is certainly not the case that deconstruction prevents us from saying this, or (to invoke the target of Gottdiener and others) that postmodernism – 'the enemy of authenticity' – allows only for 'decontextualized' meanings that would prevent us from situating ourselves in the context of an opposition to anything at all. Without certitude, anything goes!

That is emphatically not what I argue here. Nonetheless *Beyond Semiotics* is certainly against certitude, a point to which I return in the coda. What it's against is not semiotics but a certain way of thinking, called 'logocentrism' (see Chapter 5), which operates from a position of certitude in regards to the nature of identity. Identity, then, is a cardinal figure or core concept of logocentric thought (which in a sense is also to say, as we will see later, thought in general). But as our rhetorical opening serves to show, it is a good deal more slippery than its idealized 'core' status suggests. To push the point a little further, I am not suggesting that the opening to this book is only or merely 'rhetorical', or that it bears the *identity* of rhetoric. I am not saying either that rhetoric is identifiably playful or frivolous in contrast to an idea of the serious as 'philosophical' or 'political'. If it is true that a certain certitude as regards the structure of identity runs deep in logocentrism, and therefore in semiotics, how could it ever be 'merely' playful, no more than a rhetorical card trick, to try to show that structure as always already unsettled and unsettling, with a view to suggesting that things are never quite as certain as they might seem? This could, of course, be a rhetorical move. But could it ever be only that? Would it not be, or could it not be allowed, that the desire to unsettle the identity of identity, in order to make certitude less certain, is in some sense political, whatever else it might also be? Let me put this more forcefully: whatever else it might be, the desire to deconstruct identity operates from – *it is* – the love of democracy. To deconstruct identity is to love democracy! So in going beyond semiotics here I am taking a stand against certitude in the name, and for the love, of democracy. You can call me a post-political white boy if you like, but that's what I see to be at stake – I could hardly have come cleaner – in this book.

In case there remains any uncertainty on this point, however, let me repeat those stakes. Better still, let me shift their identity somewhat by putting to rest any inference that they are mine, as if somehow on my own I was inspired to see that to be against certitude is to love democracy. Here then I cite a beautiful passage by John D. Caputo, commenting on (as it were) the project of deconstruction in relation to an idea of national identity:

> The idea behind deconstruction is to deconstruct the workings of strong nation-states with powerful immigration policies, to deconstruct the rhetoric of nationalism, the politics of place, the metaphysics of native land and native tongue, of *propria* and my-ownness. The idea is to disarm the bombs, *les grenades*, of identity that nation-states build to defend themselves against the stranger, against Jews and Arabs and immigrants . . ., against all the others, all the other others, all of whom, according to an impossible formula, a formula of the impossible, are wholly other. Contrary to the claims of Derrida's more careless critics, the passion of deconstruction is deeply political, for deconstruction is a relentless, if sometimes indirect, discourse on democracy, on a democracy to come. Derrida's democracy is a radically pluralistic polity that resists the terror of an organic, ethnic, spiritual unity, of the natural, native bonds of the nation . . ., which grind to dust everything that is not a kin of the ruling kind and genus.[3]

These are strong and, for me, inspiring words. It is because (I could not come any cleaner on this) I believe them, believe what they inspire me to believe, that I am against the sort of certitude that both underwrites and parades itself in such statements as 'the traditional narratives of our culture' or 'the enemy of authenticity'. I would go so far as to say that the friend of authenticity is the enemy of democracy, and to call 'the traditional narratives of our culture' (where 'ours' is not an indigenous culture, but that of a powerful nation-state) an example of 'the terror . . . of spiritual unity'! For in every example of a community (the example of culture, say) there is a particular structure of identity, based on the structure of identity in general, which binds community members together *against others*, marking the limits between insiders (us) and outsiders (them). For a certain concept of culture, for example, an appreciation of art and literature is the identity marker of insiders, while outsiders are marked as 'media philistines', 'image junkies' or 'techno-brats'.

This may suggest that, out of love for democracy, there is a need to go beyond an assumed hierarchy of texts based on a semiotics of their imputed individual identities. Not, therefore, the question of whether literature is aesthetically, morally, spiritually or in other ways superior to television, or whether traditional painting is superior to modern art or computer graphics, but the question of *textuality*, of what might be called the limits-without-limits of texts, is what we should be asking in democracy's name. In *Beyond Semiotics*, then, 'text' refers to the sign's capacity to remain, as it were, open to interpretation, chance encounters, indeterminate associations and relationships, a capacity – which semiotics suppresses – for remaining open to the other. While this is most evident in Chapters 7 and 8, it is no less important to the book as a whole. Moreover, text is not confined to aesthetic products, traditional, postmodern or otherwise; while it needs to be seen as a re-animation or re-opening of the Saussurean sign, it is also a strategic attempt to unsettle the effects of what Derrida calls the speech/writing opposition, which I introduce in Chapter 3 and may be summarized here as a founding structure of succession, hierarchy and oppositional difference. Because, in short, speech comes before writing historically, a whole metaphysics of truth has arisen according to which (to put this very sketchily for now) the things we say to one another are taken to be more true, meaningful and authentic than the things we write to one another. So writing is postlapsarian, as it were, marking a fall away from speech, from truth. This explains why technology can be understood as a form of writing (as I discuss in Chapters 3, 4 and 5), in the sense of being seen as artificial and supplementary, a dangerous add-on to 'authentic' culture which, although it may improve certain material conditions of life, also threatens to take us away from our 'true' selves as closely bonded individuals who share a deep sense of cultural tradition and spirit. Here then is another variation on our theme: in attempting to go beyond semiotics we are attempting to go beyond the speech/writing opposition, but without thinking to escape it.

One effect of this opposition is to reduce text to identifiable written objects, usually in the form of books. Truths are spoken but texts are written, then. In rejecting this (in so far as it's possible to do so, for we could never hope to go beyond logocentrism in the sense of getting outside of it), we can say that text is what our identities are made out of, what the identity of identity is made out of, and whatever we understand by 'the world' is made out of text too – it is textual. Hence by textuality is meant something *beyond* the speech/writing opposition, something which exceeds the opposition between authenticity and artificiality or the opposition of text and not-text. Text (or textuality) does not refer to a species of fiction, an aesthetic practice or product; it does not signify an artificial construct (like a poem, an episode of *The*

X-Files or an advertisement in a magazine), since this would be to oppose it to some idea(l) of the 'authentic', the 'natural', the 'organic'. Text, or rather texts (for text is precisely what isn't singular, which defines its singularity), is a *democratic* idea. The Nazis surely told lies, committed unspeakable atrocities, did things that we should never do to others; but they didn't act out of any allegiance to textuality or from a desire to go beyond the speech/writing opposition. They acted out of certitude, from an unshakeable belief that they were right to say and do and get away with what they did. How is it possible that, barely half a century later, certitude remains the order of the day, whether as the 'naturalness' of the free market, the cultural 'right' of a nation to legislate against foreigners or its 'principled' denial of sovereignty to indigenous peoples within it (yes, I am thinking of the Australian nation here and the uncivil refusal of its present government to recognize the claim to self-determination of Aboriginal people),[4] the 'justice' of a taxation system that rewards hard effort when effort alone is scarcely a means to better living standards or improved social status, the 'truth' that competition is so natural and right that any state-sponsored institution (a hospital or education system, a public utility) must be 'artificial' and therefore has to be privatized – and so on? This is not to assert that being certain is to be a Nazi. It's to say that Nazism is irreducible to an historical event, that it cannot be explained simply as an outcome of historically determinable circumstances and certainly not as the German nation's cultural desire for, or innate proclivity towards, world domination. Yes, there were events – real historical events – that made the Third Reich possible. But to suppose that therefore Nazism is a phenomenon belonging only to history through and through, to suppose that it's not in any sense a phenomenon made possible also by *metaphysics*, by a certain way of thinking, a way that thinks to know the grounds of certitude, to know the difference between speech and writing, Aryans and Jews, authentic and postmodern culture, would be to suppose either that by keeping a close check on things we'll be able to prevent the repetition of those historical events (or at least the repetition of their previous outcome) or that somehow, in the end, Nazism is inherently 'German' and so now that Germany has been re-invented as a democracy we need never have to worry about the prospect of a Fourth Reich coming into historical being. So let's go beat the crap out of the Russians! Such was the rhetoric and the logic, of course, of the Cold War: not Germany but the Soviet Union is the enemy! But is it really any different, at a fundamental level, from a rhetoric and a logic that defines postmodernism (defines it with utter certitude) as 'the enemy of authenticity'? As a metaphysical construct, is 'the enemy of authenticity' all that different, at a fundamental level, from 'the enemies of freedom' (which could be the Viet Cong, or perhaps anyone opposed to US business interests)?

These are very heady, and very heavy, questions. Let me repeat (albeit somewhat differently, as it were) that certitude does not belong to the Third Reich, or indeed to Stalinist Russia or any other totalitarian regime, as its exclusive metaphysical property. Hence certitude is not just a historico-political phenomenon, something which happens from time to time under such-and-such circumstances. In so far as certitude is logocentric, it's endemic to thought, and in that sense (a very particular one) it is outside of history. But to say that we can't escape it, is not to say we can't work against it, or to claim the only way to work against it, is to oppose it by advocating relativism. The problem here would be the oppositional structure of the choice – certitude *or* relativism – a choice which, *as a choice*, isn't structurally different from the choice between Aryan or Jew, authentic or postmodern culture and so on. If, however, the choice between certitude and relativism were seen as textual, and not in terms of an ultimate, decontextualized or transcendental difference, then we could never say, in advance of any contextual need or pressure to say, when to be certain (and about what) and when not to be.

I hope this makes it clear that 'text' does not mean 'the book' (as I argue again in Chapter 8). For this reason the topics or objects of discussion here are fairly diverse, including films, works of literature, the technologies of radio and the telephone, sounds, cybernetics, concepts, the meaning of culture, the identity of philosophy, the nature of thought, the structure of writing and a good deal more! There is no attempt on my part to control or master the open and always potentially (if not conditionally) errant effects of this dispersal – of this text – by developing a system or a method of cultural criticism. In taking a stand against the terror and the tyranny of unities, *Beyond Semiotics* is therefore a-programmatic (or 'adestinal', as we'll see later), its principal assumption being that there is no such thing as a 'decontextualized' meaning, or a decontextualized anything. 'There is no outside the text', as Derrida has famously put it, albeit the ways in which this has been misunderstood are legion.[5] But what this statement means takes us beyond semiotics: it means that nothing 'is' outside of a context! Now this has to be understood in the radical sense of contextuality without limits, in the sense that there is nothing whatsoever 'outside' of contextuality, that there's nowhere to stand outside of contexts, outside the text. It's in this sense that contextuality goes beyond semiotics, since it is certainly the case that, for semiotics itself, the meaning of any sign is always taken to be contextual, an effect of sense-producing rules and procedures belonging to particular sign-using communities. All the same, to the extent that any community-sanctioned *system* of sense-production is taken to set *limits* to the meaning of any sign, to the extent that context (even in this qualified sense) is taken to determine the limits of meaning, then contextuality is not, for semiotics, the radically groundless (or quasi-transcendental) ground it is for deconstruction. It should go without saying, moreover, that contextuality is not, for deconstruction, a licence to say anything at all about everything under the sun. The statement 'there is no outside the text' is not philosophical code for 'anything goes'! It's an acknowledgement rather (and let me say it again – an acknowledgement given in the name of democracy) that when it comes to textuality there is no ground, no rock-solid philosophical substratum or transcendental baseline we could use or refer to in deciding a text's 'ultimate' meaning. Because semiotics continues to work with an idea of context which remains stubbornly logocentric, its idea of the sign is less radical (less democratic, in a sense) than it potentially could be. Although the *arbitrary* sign is the core unit of any semiotic analysis, in its *structure* the sign remains a fundamentally stable unit grounded in a taken-for-granted idea that everything has its limits, despite the fact that semiotics offers a radical account of those limits in terms of community-based conventions. In order to go beyond semiotics, then, we need to unsettle the idea of a unit by taking up what might be called the idea of a unit-without-unity: from the sign to the text.

In their asystematic, indeterminate, nonclosural disunities, texts are always open to misreadings and misappropriations, improper uses, errant and erroneous interpretations, chance encounters with other texts. Meanings are always set in context, that is to say, not in concrete. Even what we take for destiny is never anything more than one possible outcome or effect of chance, an effect of textuality in general. This is no less true of *Beyond Semiotics*, of course, than other texts. But I would not want you to think I've deliberately written a book in which I have purposefully tried to get everything wrong or which I am actually hoping you will misread, or that I could not care less about the chances of its being 'well received'. Far from it. I want you to love this book to pieces, to marvel over its audacity, revel in its revelries, laugh at all the jokes and encourage your friends to buy it! For who could write from anything less than a desire to be affecting, to make a difference, to create personal effects? And so, dear reader, this book is just for you. For all that it's a book about the impossibility of ever getting outside of texts; it is also the expression of an act of faith (in the name of deconstructing certitude and loving democracy) in all our singular identities ... especially yours and mine.

I give it over now expecting nothing in return but knowing that I must get something back. I could never say exactly what that might be, though of course I hope that my intentions in writing the book will be understood, even if you end up disagreeing with them. Meanings are not set in concrete, but neither are they floating about ethereally in a decontextualized zone of free association, trippy intuition or personal feeling. There is simply no contradiction, then, between saying that there's nothing outside of the text and that I wrote this text with the full intention of wanting it to mean something. But of course you can't always get what you want. It's not just the Rolling Stones who understood this, at least not in so far as it may be taken for a variant of a famous definition of semiotics which Umberto Eco gives: 'the discipline studying everything which can be used in order to lie', or to stop you from getting what you might want (or indeed, for all anyone knows, the opposite of this).[6] So while I certainly do have intentions (and who doesn't?; who could be imagined to believe that intentionality is inert, nothing but a 'fiction'?), there's no guarantee – on the contrary – that you'll 'get' them or that you won't understand the book to mean things I might never have intended. This is the risk we run in writing, the risk that textuality entails, and there's no getting away from it, no 'outside' this risk. But obviously I would prefer that you understood the book to mean some things (whether or not I intended them) rather than others. With the express intention of guiding you in the direction of that preference, let me chance the risk of saying that this is a book about writing in both (what I will call for now) a conceptual and a rhetorical sense. *Beyond Semiotics* is about writing conceptually and rhetorically, for democracy's sake. In its many dispersals, changes of direction, shifts in modality and other apparatus designed to work against the idea that there's a systematic way of moving beyond semiotics (the point is that there isn't), the book is actually meant to be enjoyable to read. As much as Saussure and Derrida loom large among its influences, however, it isn't any book of philosophy that I've tried to model my writing on. If *Beyond Semiotics* has an analogue, then for me it's the Rolling Stones' *Exile on Main St.*! I've been playing that record a lot while I've been writing this, and I've come to realize only recently that what strikes me as remarkable about it is that, despite all but two of the songs being 'originals' (mostly Jagger–Richards compositions), the whole thing sounds like an album of cover versions of songs which have been around for a long, long time indeed. I understand now that what's always drawn me to *Exile* is that it sounds so 'traditional', along with the fact that it's both all over the place musically and at the same time all of its dispersals somehow come together as one, creating a unit-without-unity that, for me and many others, constitutes just about the best racket ever made. Not forgetting Mick Jagger's 'singing', of course, or rather the fact he slurs and mumbles and moans and screeches all the way through so that you can never 'understand' more than a few fragments of the lyric in just about any of the songs. As a vocal performance, Jagger's singing on *Exile* is almost pure glossolalia, a reminder that language is irreducible to words.

So here then I run the no doubt crazy risk of saying that one of the pre-texts of *Beyond Semiotics* is *Exile on Main St.*, no less than Saussure's *Course in General Linguistics* or, among other works by Derrida, the *Grammatology*. When we're in the space of cultural criticism we're never in the space only of writing in the narrow sense; indeed it's precisely that sense of writing which we need to open up and move beyond. But first we need to go back, back to an idea of culture, to see where this came from, what it might mean and where it could lead us. Try looking into my eyes as you come with me.

(They're still blue, my darling, and forever true.)

1. The Concept of Culture

Like all concepts, the concept of culture has many applications. Some sense of the many ways in which 'culture' can be used is given by the thesaurus, where its main synonyms are listed as *breed stock, learning, civilization* and *good taste*. The entry for 'culture' itself lists as synonyms the terms *literature, the humanities, liberal education, scientific education, cultivation of the mind, acquisitions, accomplishments, proficiency, mastery*.

From this list, which is only partial, 'culture' may be said to mean something like *the best of human thoughts and deeds*, confined largely to the realm of artistic activity and tied clearly to some notion of self-development. Hence the concept of culture is being lent here a certain specificity, as opposed to what might be called the anthropological understanding of culture as *a whole way of life*.

We will come back to that understanding later. But for now let's continue with this other definition of the concept which I want to call (and there are many ways of designating it) a humanistic understanding in so far as it emphasizes notions of aesthetic value and moral purpose. This is quite a common understanding of culture in its everyday use. Aesthetic products and practices are thought to make up the domain of culture, in other words, and an appreciation of this domain is justified for its morally purposive effects. The appreciation of culture makes for a better person. In the English-speaking West the educational site par excellence of this definition of culture was, until recently, and in many cases remains, the university English department. The whole point of reading English Literature was that it would make you a better person. Again, though, it should be stressed that this definition of the concept has to be set against the anthropological understanding of culture as a whole way of life.

The evidence for claiming that the humanistic understanding centres around the English department as the exemplary site of its reproduction could scarcely be stronger than these words by George Gordon, Professor of English Literature at Oxford in the early part of the nineteenth century, taken from his inaugural lecture:

> England is sick, and ... English Literature must save it. The Churches (as I understand) having failed, and social remedies being slow, English Literature has now a triple function: still, I suppose, to delight and instruct us, but also, and above all, to save our souls and heal the State.[1]

Notwithstanding that this may be a rather extreme or wishful account of the humanistic concept of culture, it is otherwise utterly stock and, if not quite dominant, remains certainly virulent today. It also allows a more general point to be made about the concept of culture as such: for Gordon, culture (in the form of English Literature) is seen to be able to succeed at moral instruction where the Church has failed. The general point is that you cannot have a concept of culture at all until the idea of social relations is loosened from a centring in God, or for that matter in nature. The historical moment of that loosening is the occasion of a self-reflexive, or self-critical, political science that can be traced back through the work of several important European thinkers dating from the mid-seventeenth to the late-nineteenth centuries. I have in mind especially Thomas Hobbes's *Leviathan* (1651), Giambattista Vico's *New Science* (1725) and (less important in itself but exemplary of a certain understanding of culture) Hippolyte Taine's *History of English Literature* (1867).

Now before saying something about the importance of these critical thinkers and their work to the emergence of a concept of culture via the formation of a political science, it is worth repeating the general point that you cannot have a *positive* concept of culture until you get rid of a God-centred or a nature-centred understanding of human social practices and relations. From Aristotle to Descartes, culture was always something that was just 'given'; it was always just 'there', the human social order being simply the *natural* order of our species as decreed by divine or some other cosmic law. Just as leaves are green, human society was understood to be the way it was because *that* is the way it is, the way it was meant to be. There wasn't anything else to be said or to think about what we would now call 'culture'. At worst, culture was a realm of learning that had to be stripped away, at any rate for Descartes, in order to gain access to the natural realm of the Cogito or the 'true' self.[2] So the only sense in which you could have a concept of culture, in those terms, was by negation. There was no sense of culture as a positive force, a human achievement, a product of individual or social will. There was no sense at all of culture as struggle or process, as achieved by acts of human imagination or political consciousness – no sense of culture as defined by individuals or communities rather than by cosmic forces. Aristotle, for example, considered the polis (or society, in the form of the city state) to be the natural habitat of 'man'; he did not think of it as a human idea or the result of political struggle.[3]

Around the Englishman Thomas Hobbes, however, working in the mid- to late seventeenth century, the possibility of beginning even to *think* a concept of culture opens up. With Hobbes we get for the first time a theory of society as a domain of human relations established by contractual agreement among all men. For the first time we get a concept of culture as a positive force, the whole point of *Leviathan* being that the human social order rises above, triumphs over, the state of nature, which Hobbes defines as 'a warre ... of every man, against every man'.[4] Nature, for Hobbes, is the state of hedonistic men looking to fulfil their own pleasures, unrestrained by laws. Both merciless and lawless, it defines a state of negation, a state of lack. Enter, then, the human contract, upon which society is founded and by which it triumphs over nature, putting culture in the place of that originary lack. The question is how to organize and structure the conditions of that triumph, and for Hobbes the answer is simple: all that men have to do is to give up their equality under a lawless regime and give over their powers of self-expression and self-gratification to a sovereign leader in the social realm. As Michael Callon and Bruno Latour put it:

> Thus 'authorized,' the sovereign becomes the *person* who says what the others are, what they want and what they are worth, accountant of all debts, guarantor of all laws, recorder of property registers, supreme measurer of ranks, opinions, judgments and currency. In short the sovereign becomes the Leviathan: 'that Mortal God, to which we owe under the Immortal God, our peace and defense.'[5]

Here it is worth recalling Professor Gordon's inaugural lecture at Oxford in the first decade of the nineteenth century. At that time for Gordon the English (if not the European) social order was in moral decline, and so it was imperative that English Literature be used 'to save our souls and heal the State'. Gordon singled out a particular aspect of culture as being powerful enough to change the human condition. It matters not at all what forms he might have imagined that change to take; nor does it matter that we might think his proposition ridiculous now, or even, according to a certain politics, offensive. The point is that

Gordon understood culture as a *positive force* and he could not have done so without Hobbes who, centuries before, made it possible to begin to think of culture in such terms. For the very idea of the sovereign is a *cultural* achievement, and for Hobbes its greatest effect is measured as 'our peace and defense'. Culture, then, guarantees our present security and future happiness.

In light of Hobbes's fundamental importance to the development of a concept of culture, it is tempting to say that before the European Enlightenment, in relation to which his ideas were at the very least precursory, *there was no culture*. This would not be to say there was no culture in any sense before Hobbes, or that before the Enlightenment God made culture and afterwards men started making it. It would mean rather that there was no available *concept* of culture from within what might be called a world-view of human history as a divinely pre-determined and more or less unchanging condition. In a word, and crucially, culture is an utterly secular concept.

Now once you've begun to develop a concept of culture, you would not expect it to take very long before someone started to imagine a *science* of culture, proposing to analyse and measure its conditions of possibility, internal properties and so on. Not too surprisingly, then, three-quarters of a century after Hobbes's *Leviathan* appeared in England, Vico published his *New Science* in Italy. This is not to say that Vico had read Hobbes or used *Leviathan* as a model for his own ideas; it means simply that Vico had available to him – as part of his intellectual history, part of his way of thinking – a concept of culture. In fact Vico's model wasn't Hobbes; it was the work of such figures as Galileo, Bacon and Newton. His model was the 'natural' science of these thinkers, because his own goal was to produce nothing less than a *science of human society*, to achieve for 'the world of nations' what the scientists had done for 'the world of nature'. He wanted to produce 'a physics of man'. Vico went about this task by arguing that so-called 'primitive' man was not actually a childish brute at all, but could be revealed through careful 'scientific' analysis to have possessed an intrinsic 'poetic wisdom' that cast his responses to the world in the form of a 'meta-physics' of symbol, metaphor and myth. For Vico, contrary to what was commonly thought at the time, any early society's account of itself in relation to the world was not naive and absurd, since it was not meant to be taken *literally*. Such accounts were actually mature and fully formed cognitive responses to physical, spiritual and social phenomena. They were sophisticated structures of encoding a society's relations with and to the world, a means of coming to terms with reality. As Vico puts it: '[i]t follows that the first science to be learned should be mythology or the interpretation of fables; for, as we shall see, all the histories of the gentiles [pagans, pre-moderns] have their beginnings in fables.'[6] This proposition is so important to Vico's thinking that it constitutes 'the master key of this Science',[7] everything turning on 'the fact that the first gentile peoples, by a demonstrated necessity of nature, were poets who spoke in poetic characters'.[8] For Vico, then, myths are not silly stories, but *civil histories*: the 'civil histories of the first peoples, who were everywhere naturally poets'.[9] And this, for the first time, accords myth the status of crucial historical evidence:

> The civil institutions in use under such kingdoms [of primitive people] are narrated for us by poetic history in the numerous fables that deal with contests of song ... and consequently refer to heroic contests over the auspices ... Thus the satyr Marsyas ... when overcome by Apollo in a contest of song, is flayed alive by the god ... The sirens, who lull sailors to sleep with their song and then cut their throats; the Sphinx who puts riddles to travellers and slays them on their failure to find the solution; Circe, who by her enchantments turns into swine the comrades of Ulysses ...: all these portray the politics of the heroic cities. The sailors, travelers, and wanderers of these fables are the aliens, that is, the plebeians who, contending with the heroes for a share in the auspices, are vanquished in the attempt and cruelly punished.[10]

A myth provides documentary evidence therefore of the actual lived experience of a people; it's a sophisticated way of giving shape to an otherwise formless reality. Imposing an order on the world, myths give shape to it – and this *shape* is what comes to be accepted as 'natural', 'given' and 'true'.

This is a real insight on Vico's part (putting semiotics and other theories of language and communication in his debt), because it unsettles the nature/culture binary. After Vico, you don't have to worry any more over what came before culture: 'nature' is more or less whatever a particular culture happens to say that it is. Culture is not something that comes *after* nature, the case being instead closer to the other way around: nature is what comes 'after' culture; culture *produces* nature. Prior to Vico, to stretch the point only slightly, this was unthinkable.

That it is able to be thought at all, Vico might have argued, is because the shape we give the world derives from human cognition, from the mind itself. This establishes the principle of what Vico calls *verum factum*: what man recognizes as true (*verum*) and what he himself has made (*factum*) are one and the same. 'So that, if we consider the matter well,' he writes, 'poetic truth is metaphysical truth, and physical truth which is not in conformity with it should be considered false.'[11] If 'nature', in other words, does not agree with culture, then 'nature' must be wrong, since what men make is true and what they make must be cultural.

But of course it is not quite that simple. I am exaggerating to the point of warping the nature/culture relation in order to emphasize the positive value that Vico ascribes to civil society. At the same time it is important to remember that, according to Vico, truths are cultural because men make them, but also because men recognize them to be true – to which extent they're a part of nature. An example might be that it is true there's no such animal as a carnivorous cow. Men have made this truth, in so far as they have manufactured a knowledge of husbandry and a science of animal species. But it is true also in so far as carnivorous cows do not belong to nature – and if one did, it would have to be false.

According to the 'physics of man', then, men 'in a certain sense created themselves',[12] for

in the night of thick darkness enveloping the earliest antiquity, so remote from ourselves, there shines the eternal and never failing light of a truth beyond all question: that the world of civil society has certainly been made by men, and that its principles are therefore to be found within the modifications of our own human mind. Whoever reflects on this cannot but marvel that the philosophers should have bent all their energies to the study of the world of nature, which, since God made it, He alone knows; and that they should have neglected the study of the world of nations, or civil world, which, since men had made it, men could come to know.[13]

On this view man is above all a *maker*, for which (it may be interesting to note) the Greek word is in fact 'poet'.

Nevertheless, this making or shaping process is not as one-sided as may appear from the discussion so far. It turns out rather to cut both ways: not only do people produce their own social institutions in light of their own imaginings; these institutions also produce them. 'Humanity,' as one of the English translators of the third edition (1744) of *The New Science* puts it, 'is not [for Vico] a presupposition, but a consequence, an effect, a product of institution building.'[14] Humanity constructs the world as it sees it, and thereby constructs itself. This structuring process is constant, habituating human beings to a man-made world they perceive as 'natural'. Although we might be able to describe this *process* as natural, that will not allow us to designate an essential property of human nature as such, since what human beings end up becoming is not pre-determined by any model of their becoming. We are always the products of enculturation, and we are also always the producers of culture. Despite being written almost three hundred years ago, then, *The New Science* seems (as noted by Terrence Hawkes) remarkably modern:

> *The New Science* links directly with those modern schools of thought whose first premise may be said to be that human beings and human societies are not fashioned after some model or plan which exists before they do. Like the existentialists, Vico seems to argue that there is no pre-existent, 'given' human essence, no pre-determined 'human nature.' Like the Marxists, he seems to say that particular forms of humanity are determined by particular social relations and systems of human institutions.[15]

This last statement need not, of course, be confined to 'the Marxists', sounding very much as it does like the anthropological understanding of culture which I referred to previously as a whole way of life. Still the point remains that, for Vico, culture creates people, and people create culture; and there is something disarmingly familiar about this, an idea which is nearly 300 years old and at the same time continues to permeate modern thought.

Curiously, perhaps, one of the interesting things about *The New Science* is that it isn't especially predictive. Vico is more concerned to give a descriptive account of culture than with establishing a scientific method that might then be used to predict when a culture would be due for some set of changed conditions and what it might look like afterwards. On the one hand, he does claim that

> the nature of institutions is nothing but their coming into being at certain times and in certain guises. Whenever the time and guise are thus and so, such and not otherwise are the institutions that come into being.[16]

Certainly, then, he does seem to regard his 'physics of man' as a fairly exact science. But, on the other hand, he is by and large not all that interested in following up the alleged predictive powers of his theory.

A century or so after Vico, though, we find that the desire for a science of culture has grown into a kind of obsessive faith, a belief in the power of measurement, after the manner – and the success – of the nineteenth-century natural sciences. The following passage, for instance, taken from the four-volume *History of English Literature* by the Frenchman Hippolyte Taine, an important figure in nineteenth-century European positivism, is both of a piece with the concept of culture at that time and a remarkable expression of the rationalist dream:

> We have reached a new world, which is infinite, because every action which we see involves an infinite association of reasonings, emotions, sensations new and old, which have served to bring it to light, and which, like great rocks deep-seated in the ground, find in it their end and their level. This underworld is a new subject-matter, proper to the historian. If his critical education is sufficient, he can lay bare, under every detail of architecture, every stroke in a picture, every phrase in a writing, the special sensation whence detail, stroke, or phrase had issue; he is present at the drama which was enacted in the soul of artist or writer; the choice of a word, the brevity or length of a sentence, the nature of a metaphor, the accent of a verse, the development of an argument – everything is a symbol to him; while his eyes read the text, his soul and mind pursue the continuous development and the everchanging succession of the emotions and conceptions out of which the text has sprung: in short, he works out its psychology.[17]

This incredible passage resonates with Newtonian confidence and a developing positivist faith. It is in fact an attempt to do precisely for cultural science (cultural studies, as it were) what Newton did for natural science.

If Newton's famous and profoundly influential *Mathematical Principles of Natural Philosophy* (1687) were to be summed up in a single sentence, then it would surely be, as Ian Stewart writes, that '*Nature has laws, and we can find them.*'[18] Hence what Taine imagines is that all we need do is to substitute 'culture' for 'nature' in order to arrive at this corollary: *culture has laws, and we can find them.* It may well be that culture is difficult to analyse, being subject to forces which appear to be as wayward as the wind, but the point for Taine is that those forces are not themselves completely aleatory. We can determine the precise conditions of the imagination by and under which a particular work of art, any work of art, is produced *and* we can determine the precise conditions of its public reception. We can do so because Newton did it with physical phenomena; why shouldn't we be able to do it with cultural phenomena as well?

This is of course a tantalizing question. What it asks is for there to be a way of predicting, say, the ratings success of a new television pro- gramme, or ticket sales for the latest Hollywood blockbuster or even the outcome of a political election. These are all cultural phenomena; why shouldn't we be able to fashion accurate predictions out of them?

One response might be that we can ask this question at all only because there is already in place a cultural belief in the power of science to predict the behaviour of natural phenomena. Since this belief has not always been in place, it is cultural, however natural we might think it is to believe it. Because we think of this belief as natural, then perhaps it is natural enough for us to want to believe that there is a scientific method – there must be – for predicting the reception of cultural phenomena.

This might be all the more convincing if the premise were true: if it were truly the case that Newtonian science is right, and that nature does have laws and so we can find them. But what if it doesn't (exactly), and we can't (always)? There are some things that science, or Newtonian science, cannot measure. To take a common example from chaos theory: if you were to mark a single point on a flag and then let the flag flutter in the breeze for a while, there is no way of calculating the exact spatial coordinates of the marked point at any moment; no way of predicting where it will be at any given time. So Taine is wrong, and so is Bob Dylan: the answer is in fact not blowin' in the wind, or if it is we've got no way of accessing it. There is no way of predicting, with absolute precision, the behaviour of a gust of wind; yet it is precisely on the assumption that it can be predicted that Taine maintains that cultural phenomena too are based on underlying principles, 'like great rocks deep-seated in the ground', and therefore we ought to be able to uncover them.

But let us pause here for a moment, to reflect on where we've been. To date we have said that you cannot have a theory of culture which isn't secular, because it is a fundamental condition of having any theory of culture at all that the prevailing world-view is not centred on God or nature. What has to have been developed, in other words, is some sort of theory of the individual, whose own actions and determinations can be understood to produce (albeit collectively) a thing called culture.

That was our first point. Thereafter we looked at the historical moment at which just such an intellectual upheaval could be said to have occurred: namely, the European Enlightenment. Prior to this moment, though, there was the important figure of Thomas Hobbes, whose work in the mid-seventeenth century may be claimed as the beginning of a political science. Not long after, the scientific approach to culture becomes far more important in the work of Giambattista Vico during the Enlightenment itself. It becomes more important still, bordering on a religious quest, for the nineteenth-century positivists.

So why, after all this, is culture so difficult to define? Surely there *must* be laws governing its production. The problem here, however, is that those very laws would themselves have to be part of culture – where could they have come from otherwise? How could they have been formed? Unless you believe in a universe controlled by cosmic forces, so that even human social institutions and practices are pre-determined by a higher power; unless you believe that men and women do not make their own history, then you would never be able to answer the question – what are the laws that govern the production of culture? For whatever you thought to have found, as instances of those laws, could not have always already been there, sitting in place or lying in wait, *before* there was culture. So any evidence of cultural laws would have to belong to the thing itself – culture – and could not be a set of a priori ruling principles. It could be a set of ruling principles only if you went back to the model of the God-centred universe. Then you could say that God or some other cosmic deity created the laws of culture, whereupon culture was produced. If, though, you do not want to hold to this; if you actually *are* interested in cultural production as a human phenomenon, then you could well want to ask – why bother with a concept of culture at all? If there is no example of a human practice or natural object which is *not* cultural; if the domain of culture includes *everything* that human beings think and do (including how they think about and what they do with 'nature'), how could you ever describe this domain or isolate its principles of formation? Everything you could say about the domain of culture would itself belong to that domain, would itself be cultural. You could never get 'outside' of culture, then, to investigate it at a distance, and certainly not from the detached viewpoint (the classical scientific perspective) of the clinical observer. Wherever you went you would always still be 'in' culture and culture would be in you – in your habits and thoughts, desires and practices, social attitudes and relations and so on. You could never escape culture, not even by going to the moon or living on a desert island.

How then could you begin to develop a scientific or any other method of analysing something as pervasive as that? So why bother? If all your methods of investigation are always going to be part of what you are investigating, perhaps it is time to give up the search for a theory of culture. Alternatively, perhaps it's time to go looking for something else, which you might actually be able to find. Instead of searching for a *general* theory of culture, perhaps we could begin to think of investigating possible ways of accounting for cultural *differences* and *specificities*, albeit this could in no way constitute an absolute break with a general theory of culture.

The point, however, is that you would no longer have to worry over the problem of where culture comes from, or where or when it first began. You would not have to go looking for the sociological equivalent of the Big Bang. There never was such a cultural moment, certainly not one which is recoverable, and why would you want to go looking for it anyway when culture is all around you and within?

'Culture' is such a pivotal, vast and pervasive concept that you cannot even think the concept of a concept without 'doing' culture, without what you are thinking *belonging* to culture and performing it. Hence there could be no way of investigating culture in general or any culture in particular that was not always already *part* of its own problem, that wasn't always (beforehand and during) inscribed in the object that ideally (or idealistically) was held to lie outside of it. There could be no method of accounting for culture, or for *a* culture, that was not always already cultural. Any method of cultural analysis must be part of what it is setting out to discover; it has to be always already written into its own conclusions.

There is a name for the theoretical position I am putting here – poststructuralism. I will have more to say about it in coming chapters, but for now I want to establish simply that the concept of culture has to have certain historical, institutional and discursive formations in place before it can be even thought, before it can appear as *a concept*; yet these formations are not in any historical or sociological sense originary, as if there were a moment or place in time when they came suddenly into being (and this despite my earlier attributions of originality to Hobbes and Vico). Now that the concept has arrived, as it were, it is pointless to ask the question – what is culture? Instead we should be asking – what are the specific intellectual, political and other coordinates that position any *theory* of culture? What if, for example, I were to say that culture is determined by climate? This *has* been said before, and it is a perfectly legitimate theory of culture.[19] It can be used to explain, for example, the cultural differences between the cities of Sydney and London: Sydney's sun and humidity attracts people to an outdoor lifestyle, getting them out into very public places, whereas London's mist and cold induces them to find pleasure and warmth at home or in the corner pub.

That seems about as deterministic as it could get, I suppose. But even if you were to argue that climate causes culture, what would you be able to say from this about the specificities of any particular culture in terms of its political organization, media structures, institutional practices, everyday life, modes of communication, flows of information, social and economic relations and so on?[20] For surely it must be possible to comment on such things, and in a way that doesn't collapse them back into a theory of culture in general, losing sight of their specificity in the bargain.

As we have seen, such a way (if indeed there is one) could never quite be a science of culture. What then could it be? For a while there, from early in the last century until about the 1980s or so, it looked as if it could have been semiotics. But the question of the conditions by which semiotics might be regarded, or regard itself, as an exemplary theory of culture will need to be taken up in the following chapter.

2. A Short History of Semiotics

It would not be entirely wrong, though it may be provocative, to say that what used to be called semiotics is now called cultural studies. This would not be to infer that semiotics as such no longer exists – yet notice the ease by which an 'as such' can attach itself to 'semiotics' nowadays, seeming to diminish it. To say 'semiotics as such' is at the same time to say that something has been lost to semiotics, that semiotics exists today as something less than it might have been. While there is no doubt that semiotics is now well-established institutionally – through specialist journals, research centres, international conferences, professional associations and the like – it could be that its institutional well-being has been achieved at the cost of an earlier revolutionary spirit that was once the heart and soul of semiotics. Saussure himself, after all, announced that the first task of the new linguistics (the forerunner of the promised mega-study of the social life of signs in general) was to rise up against the very discipline of linguistics itself. Insisting on the scientific nature of his own approach to the study of language, which he believed would supply the principles of a semiotics to come, Saussure saw the problem with linguistics (and this can scarcely be over-emphasized) to be that it had no principles worthy of the name. Almost without exception the available approaches to the study of language before Saussure were, according to him, full of 'so many absurd notions, prejudices, mirages, and fictions' that 'the task of the linguist' (if not the exorcist) must be, 'above all else, to condemn them and to dispel them as best he can'.[1]

All the same, any suggestion that cultural studies may have taken over from semiotics today is no cause for lamenting the betrayal of Saussure's ambition. Although Saussure will feature at many points throughout this book, we should not suppose that he alone presides over everything that has been said and done in the name of semiotics.

Indeed it may even turn out that Saussure is not so much the 'father' of semiotics as the father of *semiotics as such*.

For now the issue remains whether semiotics today is in some sense a disappointment. Could it be that semiotics has let itself down, by failing to live up to an earlier promise? Or is it possible that semiotics has lost its relevance only as a name while its otherwise energizing potential has passed over to cultural studies? If semiotics had never meant anything more than Saussure's promised 'science' of signs, these questions would seem far-fetched. But they may seem perfectly reasonable if put to the sorts of claims made for semiotics by a text such as Marshall Blonsky's editorial introduction to his influential collection, *On Signs*, published in 1985.

Read today, some of Blonsky's statements make it almost impossible not to blush. 'Let us say,' he writes, 'that this discourse (semiotics in the widest sense: an intense interest in signs) has been a look at the world's misery in more fineness by far than the ways in which the misery-makers conceive their handicraft.'[2] For Blonsky, then, it is not the scientific but what may be called the *ethical* nature of semiotics that must be underscored. Precisely as an ethics, he seems to say, semiotics has no obvious use and value. 'Then who wants it? What is it worth? The discourse you will read [in the form of the essays in *On Signs*] teaches hidden things to a world organization that feels it cannot/should not tolerate semiotic knowledge.'[3]

A 'world organization'? For such a concept to belong to semiotics, rather than to mere opinion or prejudice – or to paranoia – what must its features be? Perhaps it could be argued that the concept of a world organization is a politicized variant of Saussure's *langue*, in which case the relevant feature would be its indirect (or inaccessible) but effective governance of the social and political facts of everyday life.[4] Yet while it is possible to describe *langue* neutrally as 'a collection of necessary conventions' that enable the concrete production and exchange of signs,[5] one would need to take a far more critical, or even cynical, view of the 'world organization' whose function is not at all to facilitate a general interest (sign-usage) but to guard an exclusive interest (power). Unbeknown to anyone but semioticians, then, local instances of everyday life in the late twentieth century were not the expression of cultures, communities, nations, governments, citizens – or of any other individuated *parole* form, as it were, of social organization. The fact we might believe that everyday events really are coded in ways that could be termed organic is exactly what the High Council of Power that secretly runs the world wants us to think!

At this point semiotics comes to look like the discipline (if such a name remains appropriate for what should be understood only 'in the widest sense' and only as 'an intense interest') of conspiracy studies. If so, one might wonder what happened to the promised science of signs. For surely the production of conspiracy theories can and does go on in blissful ignorance of the *langue/parole* distinction. It's not a prerequisite of paranoia that you have to have studied the *Course in General Linguistics* to become paranoid. Paranoia knows no limits, and in principle everyone is free to think that the measure of a truth lies in what others might call its absurdity.

If, nevertheless, we were to countenance the possibility of a conspiratorial organization that controls worldly events, what would count as evidence of its existence? Well, of course, everything would count as evidence. Once you've decided that real power resides in the mainstream or with the ruling class or in the hands of patriarchs, you will find the proof everywhere. If the *langue* of contemporary social relations is understood to take the form of an organization that never actually reveals itself (where are the minutes of the last meeting of all the members of the world's patriarchy, for instance?), then proof of its existence can never be more than inferential. And nothing is immune from inference:

> Seeing the world as signs able to deceive, semiotics should teach the necessity to fix onto *every* fact, even the most mundane, and ask, 'What do you mean?' as if, as the Greeks thought, meaning was in every tree and brook, as it is in our packages, ads, political slogans, the artefacts that for us have replaced nature ... But these narratives don't talk candidly or explicitly. They talk out of the side of the mouth, saying something else. In the key word of the discipline: they 'naturalize' their messages.[6]

There is something decidedly fuzzy about this passage, which I think has to do with a certain conceptual imprecision. To put it semiotically, the passage seems to rely on the presumption of a certain (albeit quite odd) paradigm consisting of the following elements: *sign, fact, meaning, narrative, talk, message*.[7] Let us put this to the test by re-phrasing certain statements and then rewriting the passage accordingly:

- the world as talk able to deceive
- semiotics should teach the necessity to fix onto *every* meaning
- if, as the Greeks thought, narrative was in every tree and brook

- messages for us have replaced nature
- these facts don't talk candidly or explicitly

Seeing the world as talk able to deceive, semiotics should teach the necessity to fix onto *every* meaning, even the most mundane, and ask, 'What do you mean?' as if, as the Greeks thought, narrative was in every tree and brook, as it is in our packages, ads, political slogans, the messages that for us have replaced nature. But these facts don't talk candidly or explicitly. They talk out of the side of the mouth, saying something else. In the key word of the discipline: they 'naturalize' their messages.

There would appear to be little substantive difference between these two texts – between, as it were, the authentic and the apocryphal version of the same. Surely this would not be the case, however, if the apocryphal version were to begin, 'Seeing the world as signs [for which substitute any of the following: facts, meanings, narrative, talk, messages] able to assist'. And what this may indicate is that, for Blonsky, semiotics is more a matter of attitude (which for now may be allowed to stand in for 'ethics') than a mode of analysis. Semiotics is an ethics rather than a science, in other words. By the same token, Blonsky could never let attitude alone become a sufficient condition of semiotics. His 'semiotics' is proudly attitudinal, but – and this is crucial – it is never not scientific. It is never, at any rate, able to let itself be seen as unscientific. If attitude alone were allowed to define semiotics, what would separate the semiotician from the man in the street?

Blonsky's problem, though, is that he wants the semiotician to be both an everyman and a specialist (if not quite a scientist). For him semiotics begins from a certain attitude towards the world, a world seen to be full of things which are capable of 'deception'. Whether those things are called 'signs' or 'facts' or 'narrative' or whatever is irrelevant. But it's absolutely fundamental that they are seen to be able *to deceive* rather than, say, to assist or empower, etc. The attitude of the Blonskian semiotician is such that signs are out to trick us – and this may be *so* far from Saussure's understanding of the sign as not even to warrant the name 'semiotics' as a mark of its distinction from 'lay' versions of what might be called the same attitude. For Blonsky it is very much to the point that no one should have to undergo specialist training in order to know that things are not what they seem, yet of course it would spell the end of his version of semiotics if in fact everyone did 'know' this! If it were common knowledge that signs are not to be trusted, what need of semiotics? So Blonsky's version of semiotics requires both the attitude (to signs) *and* the science (of signs). Of course without the right attitude (namely that signs are not to be trusted), the mere science of signs would be semiotics in name only – in a word it would be only 'semiotics' *as such.*

Now to be sure, something that we might call Saussure's attitude cannot have been irrelevant to the fact that he wanted to re-invent the discipline of linguistics. On the evidence of the *Course*, however, Saussure seemed to think it was the science of the new linguistics rather than the attitude of the new linguists that would stand as the measure of his ideas. Seventy years later, though, Blonsky's 'bias' (as expressed in his 'Introduction' to *On Signs*) is of such an order that Saussure appears as a fairly minor figure in the history of semiotics. Saussure's importance rests, for Blonsky, on his recognition that the sign 'is not substance, it is the correlation of two sets of differences'.[8] This is an important insight but one that is ultimately too formalistic, according to Blonsky, since it fails to account for what always remains unformed in every sign – its thingness, as it were: every sign is 'also a thing'.[9] This may suggest that Blonsky takes semiotics not for a scientific method, but as a way of life based on knowing that all signs (even, perhaps, especially, those containing 'hidden things') are able to be observed and analysed, and in this knowledge one is empowered to resist those signs that try to 'deceive' and so to rob us of our 'nature'. Such knowledge would count as 'the science of signs' in the way that a knowledge of the natural laws of the universe counts as the science of physics (see Chapter 1).

Again, though, Blonsky's semiotics is less a science than an ethics (something that for now might be termed a way of life). 'Moral reaction,' he writes, 'with the understanding gained through semiotics, is up to us now. We must read, understand the sign, the discourse of images, and react.'[10] For him, what he calls 'the semiotic instrument' is not designed simply for purposes of accurate description (as might be true of scientific instruments): '[w]e need to, and can do things other than watch signs make and *un*make their meanings.'[11] Semiotics is not just for *looking at* signs scientifically; on the contrary, the very fact that we do look at signs is a serious problem for Blonsky. It isn't signs we should be looking at, but ourselves. So the whole point of semiotics, on Blonsky's account, is to get us to see that we can never know who we truly are until we can see through the deceptions of a corporatized, sanitized, mediatized and ultimately *Americanized* world of signs. In the end, semiotics gets down to being the People's hope for overthrowing the forces of US imperialism! And today the most imperial

power of all (true source of the US/world organization's dominance) is
the image:

> The US is an image country in which image is much more
> important than social or economic facts. Reagan has been the
> epitome of this image power. He is the pleasure president in a
> consumer pleasureland. He lives a short workday and has a long
> fun day. He is the man of the moment, standing for American
> resurgence. He may indeed be relying too heavily on the power of
> his image and too little on the power of economic and world issues.
> The interest rate harms the country; the cold war, the world; and
> Europe, the victim of both, suffers crises. America seems too
> powerful and too dangerous to Europeans.[12]

Blonsky is right to want to draw attention to the semiotic importance
of nonlinguistic signs (a point to which I will return again and again).[13]
What he says about Reagan, though, owes more to a certain brand of
liberalism than to semiotics (or to semiotics-as-such). Anti-Reaganism
was the ideological default setting of every liberal in the Western world
in the mid-1980s, and that too had nothing directly to do with the *Course
in General Linguistics*. For what role is played by Ferdinand de Saussure
in statements such as 'Reagan stands for American resurgence' or
'Reagan relies too much on images and too little on economics'? Without wanting to contradict the claim that Saussure is not the touchstone
of all semiotic authority, would we want to say that something
belonged to semiotics if it had absolutely nothing whatsoever to do
with Saussure?

This last question raises the spectre of semiotics' disciplinarity, which
we will come back to in the coda. Meanwhile a more pressing question
remains: what need of semiotics in order to say that I'm opposed to the
Ronald Reagans of the world (which is to say, *I am a liberal*) or that I
think that 'images' are dangerous (again, *I am a liberal*)? To return this to
its source – what does the word 'image' mean (or what does it signify) in
the passage from Blonsky above?

Clearly 'image' is intended not to be met with approval. More than
this, it triggers a very certain response or reading. To be on the side of
the image one has to be against ethics! Now there is a sense in which to
be 'against ethics' is to be as close as possible to an affirmation of ethics
(and therefore to be ethical) in so far as such an affirmation remains
faithful to the nonpositive positivity of ethics. But the argument that
supports this sense is a difficult one, and it risks enormous misunderstanding.[14] Here, though, 'ethics' may be glossed as code for 'sensitiv-

ity'. In a word one shows (to oneself and others) that one is ethical by being *expressive*. That is to say – and the extracts cited from Blonsky so far are evidence of this – one deliberately (or self-consciously, etc.) produces or presents oneself as 'ethical' according to the rules of a certain grammar (or genre, or discourse, etc.) which demand that imprecision is the very measure of sensitivity.

This is to oppose semiotics and science, but also ethics and science. The ethical semiotician is opposed to the scientific scientist. In other words science is so manifestly 'bad' that no special term is needed to designate what is 'wrong' with the scientist: what is wrong with the scientist is that he or she is 'scientific'. Due to the Saussurean tradition of semiotics-as-such, however, semiotics is not manifestly but only latently 'good': hence the need to mark the nature of that goodness with a special term – ethics. The semiotician is ethical; the scientist is scientific. What this means, of course, is that the scientist is 'unethical', but there's no need to say this in so many words.

Such a reading, however much it may be 'against' Blonsky, is utterly faithful to Blonsky's way of reading as evidenced in the 'Introduction' to *On Signs,* a way of reading belonging to the kind of semiotics which he says is 'part of the bloodstream' rather than the kind (semiotics-as-such) which is merely 'added to the world, an afterthought from a monastery playing its part in leadership, in a power élite that has created an image – America the Beautifully Bland'.[15] It is faithful to Blonsky's version of semiotics because it takes its object to be something which is unsaid, or in other ways unmarked. Blonsky does not say that scientists are unethical, nor indeed that semioticians are ethical. But what he does say (pretty much in so many words) is that 'images' – by which he means advertisements, TV newsreel footage, photographs in magazines and so on – are calculated to deceive. So the object of semiotics ought to be the ethics of image signification rather than the science of image production. After Saussure, it is possible to describe 'scientifically' how an image is produced (according to the principle of differential rela-tions). But that would not be enough for what could be called an organic semiotics 'of the bloodstream', which would want to ask after the ethics of what images signify. According to this distinction, production is able to be described; signification has to be judged.

And – according to this distinction again – judgement holds sway ethically over description. Judgement is bold; description is ... bland, monastic, supplementary – in a word, American. But if it's 'American' to describe, what is it to judge? And of course the answer would seem to be that it is *European* to judge! 'Europe', for Blonsky, is 'the victim', and

for him the value of semiotics is its unique capacity to reveal 'the
world's misery'. On this account, 'the world' has a single referent –
Europe. So for Blonsky the value of semiotics is that it takes its object to
be the revelation of *European* misery, brought about by *America*. The
world is European; what is not of this world is 'American'. Europe is the
world; America is its other, from outer space.

The Old World (culture) is the world; the New World (image) is
added on, merely 'an afterthought'. First there was Europe, then came
America. First there was culture, then came the addition of images: from
authentic to postmodern culture, as it were.

But what does 'culture' mean? Here its meaning is clear: whatever is
put at risk by 'images'. Culture is European; images are American. From
this, could it be said that images are American and (by comparison,
what else but) *language* is European? Ergo: culture is language? Isn't this
precisely the basis on which Blonsky attacks 'America': America is
image rich, hence it is language poor? Blonsky's 'bland America' is a
culture in name only – only a culture as such.

What could underlie such a prejudice, turning America into a
despised instance of *inauthentic* culture? How did cultural 'authenticity'
get to be associated exclusively with Europe? In responding to this we
will need to go looking, in the following chapter, for the grounds on
which a notion of authenticity can be naturalized and therefore admir-
ingly, if not self-righteously, opposed to an ideal of the 'inauthentic' in
the form of *technology*, which would seem to be the real (if unstated)
target of Blonsky's attack on the 'Americanization' of culture. To this
extent there is nothing new in Blonsky's attack, nor should his nostalgic
longing for those 'better' times when we were all so much closer to our
'true' selves come as a surprise: as Derrida notes, the 'denunciation
of technology in the name of an originary authenticity' has a very
long history indeed, 'its logic seemingly uninterrupted from Plato to
Heidegger'.[16]

3. Total Eclipse of the Heart (Thinking through Technology)

Not for the first time, I want to underscore the importance of the speech/writing opposition as Derrida identifies it in the work of Plato. In the *Phaedrus* Plato recounts a fable whose moral is the bad effects of writing, a moral deriving from the choice he makes in thinking to resolve the dilemma that writing poses.[1] Hence Plato understands writing in terms of the *pharmakon*, a term translators have always taken to mean either 'poison' or 'cure'. But, for Derrida, Plato's moral can make sense only by overlooking the Greek word's radical undecidability. In condemning writing, then, for seeming to threaten to take the place of 'real' memory or the 'living' speech of the mind, it can be seen that 'what Plato *dreams* of is a memory with no sign. That is, with no supplement.'[2] Yet the constitution of this 'pure' memorial speech turns out to depend, according to Plato, on another form of memory, *hypomnésis*, thus compromising or contaminating the very notion of 'pure' speech as such. If speech is always already a form of supplementarity, then it could never be in a relation of absolute opposition to writing as commonly understood in terms of a technology for copying speech and therefore as the very model of supplementarity itself.

What it means, of course, to say that writing is a *technology* for copying speech must depend to some extent on what the word 'technology' means. The dictionary gives five senses:

1. *The branch of knowledge that deals with industrial arts, applied science, engineering, etc.*
    So technology is a branch of knowledge, like semiotics, dealing with a particular object or set of objects.
2. *The application of knowledge for practical ends, as in a particular field, e.g. educational technology.*
    So we could speak of semiotic technology, say, in terms of the

practical application of semiotic methods and approaches to the analysis of a text or object.

3.   *The terminology of an art, science, etc.; technical nomenclature.*
     The technology of semiotics is its set of specialist terms, its terminology.

4.   *A technological process, invention, method, or the like.*
     So: computer technology, solar technology, electrical technology ... or the technology of writing.

5.   *The sum of the ways in which a social group provide themselves with the material objects of their civilization.*
     Here we might refer to Western culture's so-called technological determinism, its single-minded commitment to the proliferation of material objects produced by technological means.

Now we don't usually use the word 'technology' to refer to a branch of knowledge, the application of a knowledge or to describe a terminology. So it's the last two senses above in which the word is most commonly used: to refer to a process, an invention or a method that produces something material. More often than not, what is understood to be produced by technology is a machine, something mechanical, like a car, a computer, a washing machine or a weapon, rather than, say, a bar of chocolate, even though a chocolate bar is certainly the result of a technological process.

Nowadays, moreover, the word 'technology' has come to resemble some of its products; so that now when we think of technology we tend to think of computers, nuclear weapons and electronic communications media hardware and software. At any rate we tend to think of these products as symbols or indices of technology, much more than we do, say, of a washing machine or a car.[3]

What then do these products – computers, nuclear weapons and the television set or a telephone – have in common, because they would seem to narrate a set of family resemblances collected by the term 'technology'? Well, of course, they don't have to have a *single* feature in common, any more than members of the same family all have to have the same eyes or nose in order to belong to that family, but it's nonetheless interesting that they seem to have had attributed to them the quality of being beyond our ken in terms of how they work, a quality that can easily slip into a fear of them being, potentially at least, out of our control.

What may be interesting about this is that it seems to be not so much intrinsic properties of, but attitudes to, technology that define the term

– social attitudes, as it were, which seem to derive their fear from a lack of understanding about *how technology works*. We don't understand how a computer works, so we call it 'technology' – just as we don't understand how a television screen can present us with an image of someone who is not actually there in the room with us, or how a telephone can allow us to speak with someone else who might be on the other side of the world (see Chapter 4). It's like magic, in other words, a kind of alchemy come true.

But, if you think about it for long enough, even a washing machine seems magical – or a fridge or a car. Does anybody really understand how you can pour some fluid into a lump of metal on wheels and make it go faster than a horse? I certainly don't, although I sort of understand the principles of combustion.

The point is that what we now think of as technology, what we *call* 'technology', in a sense has not yet fully arrived in the social present; it hasn't been fully present-ified, as yet, but seems almost to belong to the future, or to provide a glimpse of the future, and thus to be caught up in a kind of future tense, or at least a futuristic tense. It doesn't seem to be *of* this world but to have arrived out of nowhere, from someplace else.

But, of course, it didn't. So why is our present understanding of technology so liable to the kind of account I've just given? Here it is important to remember that earlier technologies have also been met with fear, suspicion and sometimes derision for the 'pre'-technological pasts they are imagined to replace (as we will see in the following chapter). People expressed serious misgivings about the imagined social effects of the steam engine, electric light, the telephone, the motor car, the aeroplane and so on long before anyone got freaked out about computers and virtual reality. Always the fear expressed is for a nostalgic 'loss' – an imagined set of 'natural' or 'authentic' community relations that will be lost to technological innovation.

I think a perfect (albeit, perhaps, unexpected) example of the narrative I am describing – technology as representing an inevitable future that forgets a better past – is John Wayne's last film, *The Shootist*, one of the screen's great Westerns. In the opening few minutes of this film the John Wayne character, a metonym of dozens of other performances that semiotize his performance in *this* film, rides in from the cinematic West (America past, represented by the wide open landscape and the threat of death) into a glimpse of future America in the form of a becoming-technological town (electricity poles feature prominently at this point), a town and a future that *must forget* its past.

I have deliberately not chosen a science fiction film to illustrate the point I wish to make, which is that one popular understanding of technology is that it brings about change at the expense of continuity; that it brings in the new on the basis only of an act of forgetting – forgetting the past. Each new technology, then, appears to have no history, as if it came out of nowhere, the product of its own production – an act of autogenesis. I have chosen not to illustrate this by way of reference to a science fiction text because I do not want technology to be thought of in terms only of sci-fi images. Nevertheless it is important to remember that *The Shootist* (directed by Don Siegel) was made in 1970 and can be read therefore as a kind of fable of our times with respect to its position on the alleged memory loss that technology brings about, a position taken up in the nuclear age. So the film's electricity poles could be read as an index of the film's diegetic future and our present.[4] In any case, within the diegetic world of the film they stand for a present which is disconnected from the past, except for the embarrassing reminder of the John Wayne character who carries history with him into the town – the town's pre-history. The montage of clips from Wayne's earlier movies at the beginning of the film thus constructs a legend – a powerful index of the town's condition of possibility, which it must forget. The problem, then, is that the legend is still alive, connecting the new techno-America of the future, the America of progress, even the nuclear America, to a *living* past, a living genealogy (which, clearly, is patriarchal, but that's another matter). So the solution must be *to kill* the John Wayne character and the history, the continuity, he embodies, even though the character (and Wayne himself) is in fact dying of cancer, with only a few weeks to live in the film and not much longer in life.

From this account of the film it may be noticed that the future is presented as always already inevitable, that the power to determine the future is invested in the techno-world. John Wayne can't win, in other words – nobody can, against what Zoë Sofia calls the '*collapsed future tense*' of nuclear technology, or simply technology in general.[5] To use a popular expression: you can't fight progress. That is to say, the future is already written: it is here, in the present, and you can't turn back the clock. Such is the discourse of technology, and it's one that even John Wayne cannot overcome. The best the film can offer is for the inevitable techno-future to retain at least a trace of continuity with the past. That's the only hope there is, small and compromised as it must also be. And that hope is carried by the boy, played by Ron Howard, who develops a close relationship with John Wayne after initially rejecting him when Wayne rides into town at the beginning of the film. In the course of their relationship, the boy effectively gets a history lesson from the older man; so that when Wayne is finally gunned down (yes, in a saloon) the boy is alive *to remember*. History thus lives on: the lineage (a patriarchal genealogy) continues, only just linking the past to a precarious present into which the future has been collapsed.

One version of that future is, of course, an imagined one as represented frequently in science fiction cinema, most notably in Stanley Kubrick's *2001: A Space Odyssey*. But before turning to a brief discussion of that film, I want to draw attention to the similarity between my account of *The Shootist* and my account of Derrida's account of Plato's account of the bad effects of writing, in the form of the fable I referred to at the beginning of this chapter. According to that fable, writing is bad because it threatens to bring about a loss of memory, thereby threatening to replace history. In relation to my remarks on *The Shootist*, then, it's as if Plato's fable were a kind of pre-text of that film. The film's celebration of human memory, in other words, and its critique of technology is another form of the speech/writing opposition. In this sense, Plato's fable is its pre-text. Yet there needs to be in the world, as it were, a certain kind of technology already in place for the association between it and the bad effects of writing to make sense. And the kind of technology I have in mind is the kind I referred to previously – whose workings are beyond our ken. This could mean any technology that 'arrives' in the world after we do, so that there is no community memory of its being always there already, as *part* of the world. It might as well have come from outer space, then; it might as well be magic. This is certainly true, I think, of the qualities associated with modern communications technology, especially space technology. Like writing, its effects cannot ever be fully controlled by whoever is *doing* the writing or working with the technology. Hence the classic anthropomorphic slip: technology has a 'mind' of its own.

So I think this slippage needs to be in place, in the world, in relation to technology (and by association with writing, or vice versa) for the critique of technology offered by *The Shootist* to make sense. For this reason, I think another of *The Shootist*'s pre-texts is in fact the film *2001: A Space Odyssey*, released in 1968. In Kubrick's film, technology achieves the status of a character in the form of HAL 9000, the ship's computer. In *2001*, then, technology develops into an icon, process and thing having become inseparable – precisely the kind of future which is indexed in *The Shootist*.[6] The problem here, of course, is that technology has developed its own history, which is independent of *human* history and threatens to obliterate it.

This is best illustrated in the long scene in which HAL begins to act out of character, as it were, no longer taking orders but giving them. Instead of responding to commands, he begins to assert his own *will*. He shuts down the human crew's life-support systems, having just set adrift in space one of the only two human crew members awake on board the ship. The other, David Bowman, or simply 'Dave' as HAL calls him, sets out in a pod to rescue his comrade, or at any rate to recover his body, only to realize that HAL will not cooperate in Dave's re-entry. Ditching his crewmate's corpse, Dave is forced to effect a dangerous means of getting back on board and then proceeds to dismantle HAL who, in the course of being 'killed', narrates a kind of life story in which he identifies the scientist who created him and sings a song, 'Daisy', that his father–creator taught him long ago (in HAL's childhood, one is tempted to say).

There is, I think, something very moving about HAL's 'death', the tragedy of which he expresses in terms of a loss of memory ('I'm losing my mind, Dave') – the memory of his origin. In other words, HAL *has* a history, and this is what lends him credibility as a character. It's what anthropomorphizes him: the fact that he has a history *and* that he remembers it, these two aspects being inseparable. Indeed, in the end HAL has far more 'personality' than Dave Bowman, who is truly machine-like in his efficiency and so emotion-less as to remain perfectly cool, calm and collected in a terrifying crisis. Hence it would seem, against the film's intentions, as it were, that computer intelligence does triumph over human emotion; but the problem is that it's the human character who triumphs, a human who has become more machine-like than a machine. Once again, then, as in *The Shootist*, technology is the winner: it's replaced human history with a history of its own, making obsolete the kind of sentimental (if also very 'manly') values that John Wayne represents in the Siegel film. There's no room for that kind of history any more, a sentimental history, but only for a programmed future, which is already present, of ever-increasing progress, precision and profit.

These are the 'shiny goods', as Sofia calls them, of technology's promise. In *2001* (again, to use her term) they are indistinguishable from 'slimy bads'. But this is not how Sofia herself sees it in her discussion of the film. While it's true that, after HAL's memory has been shut down, 'we're left', as she puts it, 'with an intact spaceship of the same genre as the nuclear bomb', this does not have to mean that 'the film sustains *without resolving* the conflict between good and bad tools'.[7] For the spaceship (and by association, the bomb) is now in the hands of a human being whom no *sentimental* reader or audience member could desire to want to become, precisely because Dave Bowman *lacks* any sentiment, just like a machine. So it's HAL who gets our pity, because he remembers where he came from – his *pain* is that he knows he's about to forget. Dave, on the other hand, appears to have come out of nowhere. He might *look* like us, but that is cause only to recoil from him: because there is nothing to *connect* him to us. Between us and Dave Bowman there must have been a massive break in human history, a huge rupture with the past, a total eclipse of the heart. That's where the film's ethics, then, might be seen to lie – in getting us to prevent that loss from happening.

My problem, though, is that this 'loss' can be only an imagined one, because it assumes there to have been a pre-technological origin from which human history is in danger of radically departing, of radically losing contact with it. It is important to see technologies, as Sofia points out, in terms of discourses that can be understood as masculine reproductive fantasies (it's surely no accident that HAL is the creation of a male scientist). But it's also important to see discourses on technology as a form of Plato's fable of the bad effects of writing, a moral able to be pointed on the assumption only that there's a more reliable, natural and original form of communication or way of remembering human history: namely, speech.

What the distinction between speech and writing doesn't see is the very ground of that distinction – what makes the distinction possible in the first place. It is impossible, in other words, to think of the difference between speech and writing as the difference between an absolute correspondence of word and thing (speech), on the one hand, and an imitation of that 'pure' or 'first' condition by means of a structure of differential marks (writing), on the other, without that *structure* coming first. If 'writing' is the name of that structure, as it is for Derrida, then writing is the originary condition of speech, or speech is an effect of writing. Hence truths and origins are effects of writing, too – products of human semiosis.

It's for this reason that we cannot resolve any questions concerning technology if we see technology, on the one hand, as representing a swerve away from human nature, human history, in the form of a more authentic past and, on the other, as a form of autogenesis, without any history at all. Neither the Platonic nor the scientific myths of technology (as the two sides might be called) will do. There is no question that the nuclear question cannot be left simply to the military and scientific communities to decide; there is no question that the nuclear question, if left unresolved, threatens to put an end to semiosis once and for all – by wiping us all out, and stopping history dead.

But neither is there any question of preventing this from happening (as would appear, from the previous chapter, to be Blonsky's position) by pretending that technology has only just arrived (whether from the States or outer space), that it's only just begun and therefore threatens to take us away from ourselves, to somewhere we've never been. Instead, technology is inseparable from human history – that is how we have recorded it, passed it on, kept it going.

That's where we need to begin – even (as we will see in the following two chapters) when the technology in question is something as mundane as the telephone or the radio. In these chapters, too, we will move still further from a straightforwardly semiotic approach to culture in favour of a post-structuralist (or deconstructive) mode of inquiry, with a view to following the implications (the entanglements, aporias and occasional blind alleys) of the speech/writing opposition. To this end we will become increasingly less reliant on a standard set of terms and concepts associated with semiotics-as-such. In seeking to move 'beyond semiotics', however (which is to say at the same time in seeking to move towards deconstruction), we could never hope to leave semiotics behind entirely – as if deconstruction were not itself indebted to Saussure's concept of the sign; as if post-structuralism bore no relation to structuralism; as if somehow the groundless ground of the speech/writing opposition were an idea that just happened to fall out of the sky!

4. The Phake Fone: Crossing (Telecommunication) Lines

tele-[1] (Gk) 'distant,' esp. 'transmission over a distance.'
tele-[2] (Gk) 'end,' 'complete.'
phono- (Gk) 'sound,' 'voice.'

telephone = distant sound or voice, or the transmission of sound or voice over distance.

Although 'the' telephone was invented in 1872, the word 'telephone' was coined in 1840, seven years before the telephone's 'inventor', Alexander Graham Bell, was born.[1] Word and thing are thus separated by a distance of thirty-two years, as if the telephone had left an impression of itself prior to its arrival – like a message on an answering machine.

It is tempting to speculate, then, that the telephone was willed into being simply by the act of naming it. Before it could become 'real' (and this might apply to all technologies, without precluding the effects of accidents or chance encounters), the telephone had to be *imagined*. Indeed, history seems to support this view: Bell's patent was lodged only a couple of hours before an attempt by Elisha Gray, who was himself but another of many inventors in the 1870s working to achieve the means of distant voice transmission.[2] So it would appear that such means were not only imaginable but had in fact been imagined, and not by Alexander Graham Bell alone, although it is with his *name* alone that the achievement of those means is now associated, despite even this fact being far from true.

But even if technologies are, as it were, imagined into existence, this need not suppose that their immediate or eventual uses will accord with their imagined ones. It does not follow that technological inventions are inscribed with natural or pre-given social, technical, political or other uses. So it was that on 15 March 1877, the year in which the world's first public telephone service was introduced, the New York *Daily Graphic* carried a front-page headlined warning, 'The Terrors of the Telephone – The Orator of the Future'. 'Curiously,' as Marshall McLuhan remarks, 'the newspaper of that time saw the telephone as a rival to the press as P.A. system, such as radio was in fact to be fifty years later.'[3] McLuhan's curiosity derives from the *Daily Graphic*'s depiction of '[a] disheveled Svengali stand[ing] before a microphone haranguing in a studio. The same mike is shown in London, San Francisco, on the Prairies, and in Dublin.'[4] It would thus seem, and it's only in retrospect that we may find it odd, that the newspaper had no understanding of what the telephone was *for*. This, quite simply, is because what it was 'for' was the transmission of sound or voice over distance, which naturally could take many forms but not all or even any of which were then imaginable. Of course there is no intrinsic reason why the telephone could not have developed into an instrument of public address, providing orator–callers with the means of spreading, say, political messages across distances beyond the range of the human voice. Neither is there any intrinsic reason why the telephone should not have been, as the *Daily Graphic* seems to have assumed it was, a monologic rather than a dialogic device.

The point is that the telephone, like other technologies, is not the product only of the history of its own invention as a technical device or object; it is also, in its present form, the product of a *social* history of its use-designation. Similarly, the invention of radio technology (as we will see in the next chapter) did not immediately coincide with the formation of a broadcast institution – it took about thirty years for radio as we now know it to develop the appearance of its present use. To this extent, telephony is an instance of what might be called technology-in-general, though it's mainly in terms of its specificities that I want to consider it here. At the risk of too many repetitions, however, I want to stress again that whatever might be said to specify the telephone needs to be understood not simply in terms of the Kantian noumenal (the thing-in-itself), the telephone as 'pure' object, but within a context of its social uses, regulatory controls and everyday practices.[5] Not simply the telephone as 'pure' object, then, but rather the telephone as a *policy* object.

By the same token, returning for a moment to the telephone's imagined history, the social history of the telephone's policy objectification is underpinned by a desire to overcome communication barriers which can be traced to its official inventor, if not also its forgotten ones. As McLuhan puts it:

> The invention of the telephone was an incident in the larger effort of the past century [the nineteenth] to render speech visible. Melville Bell, the father of Alexander Graham Bell, spent his life devising a universal alphabet that he published in 1867 under the title *Visible Speech*. Besides the aim to make all the languages of the world immediately present to each other in a simple visual form, the Bells, father and son, were much concerned to improve the state of the deaf. Visible speech seemed to promise immediate means of release for the deaf from their prison. Their struggle to perfect visible speech for the deaf led the Bells to a study of the new electrical devices that yielded the telephone. In much the same way, the Braille system of dots-for-letters had begun as a means of reading military messages in darkness, then was transferred to music, and finally to reading for the blind. Letters had been codified as dots for the fingers long before the Morse Code was developed for tele-graph use. And it is relevant to note how electric technology, in like manner, had converged on the world of speech and language, from the beginning of electricity. That which had been the first great extension of our central nervous system – the mass media of the spoken word – was soon wedded to the second great extension of the central nervous system – electric technology.[6]

It is not only the lesson of how technologies can end up being used for purposes originally unforeseen that I want to emphasize here. I want to draw attention also to McLuhan's claim of the technological effort to 'render speech visible' in the nineteenth century. In a general sense, this effort coincides with a desire to overcome communication barriers by developing not simply a universal 'language', but specifically a universal *writing*. The ideal form of such a writing would be a transcendental (or 'decontextualized') system of pictogrammatic signs, based presumably on an assumption that the different languages of the world are simply different sets of material signifiers of the same universal signifieds. The desire to convert words to pictures, then, might be indistinguishable from the desire to democratize speech; for if indeed there were such a thing as a set of universal signifieds, then it should be possible to fully represent them pictorially and thus to constitute even the deaf in terms of 'speaking' subjects. Moreover, a shared set of pictograms would enable people everywhere, from all language communities, to 'speak' to one another. Hence it would seem to be an in-principle of pictorial writing that it can overcome two obstacles, two communication barriers, of spoken language: 1) the ability to hear and 2) the necessity to learn a set of rules, which differ from language to language. Visible speech in other words would permit the voice, translated into pictures, to be heard always already, collapsing distances and differences. As McLuhan writes, the Bells' intention was 'to make all the languages of the world immediately present to each other in a simple visual form'. Nor should a desire for a universal alphabet of pictures seem far-fetched or simply quaint, as we need consider only the rhetoric of the early part of the last century surrounding the capacity of silent cinema to reach and 'speak' to audiences of all cultural kinds and language types.

Since the effort to derive a visible speech is driven by a desire to overcome distance (and therefore difference), it should not come as a complete surprise that it led to the invention of the telephone. This is not to suggest that the telephone enables one to speak in tongues, as it were, though it has certainly been understood in policy terms as an instrument of democratization and, technologically, it certainly lends speech a power to overcome the tyranny of distance. But what this presupposes is that speech, in what might be termed its 'natural' form, is a perfect medium of information exchange, so that its invention represents what McLuhan calls 'the first great extension of our central nervous system'. Whether speech in fact comes before writing, as McLuhan clearly seems to suppose, is a matter of some considerable conjecture, certainly philosophically, that we needn't go back over here (see Chapter 3). Instead, taking McLuhan's proposition as it stands, we can see that the limitation of speech in its natural form is that it can extend a speaker's 'central nervous system' beyond the boundaries of that system only in the physical presence of the speaker him- or herself. The *physical* self must always be present, then, for the psychic, emotive or some other condition of the self to extend, by means of natural speech, beyond the physical, material or actual coordinates of that condition. Quite simply if the only means of self-extension available to me were natural speech, then I would *have* to be physically present here and now to present this chapter. I could not have delivered it from the university where I work because, unless you were near enough by, you would not be able to hear; and I could not, an hour ago, have sat where you are sitting now and delivered it, because spoken words don't leave behind traces of themselves in the places where they were spoken. Yet consider this: why couldn't we have all stayed at home, our phones connected to a link-up, and I could have given and you could have heard this chapter telephonically? Or I could have phoned through the chapter from my office and you could have listened to it through a PA system in your living room. Or I could have made a cassette recording of this chapter and had it played at your place in my absence – and so on.

Now I don't wish to suggest that my presence at the writing of this chapter doesn't (or didn't) constitute a specific form of communication practice with its own specific effects, which are doubtless different from those attending other forms. But it is clear that my physical presence is not, right now, as you read (and as I write, or wrote) a *necessary* condition of this chapter's being what it is. The most that could be said is that my being here is (or my being there was) necessary only to a particular occasion, a specific form of enactment. Clearly, though, as *writing*, this text may be enacted through other forms, none of which depends on my physical presence, nor even on yours, at the time of enactment.[7]

There are two important points worth making here. First, if we overlook the question of writing as commonly understood in terms of a system or technology for copying speech, then none of the other forms in which this chapter might have been presented is without a specific history that we can afford to overlook. Until recently, in other words, one could not have contemplated presenting this chapter telephonically or, more recently, on cassette tape or, more recently still, on the internet. In general, the specific histories of the different technological means of recording or relaying the human voice constitute what McLuhan calls 'the second great extension of the central nervous system'. So it would seem that electric or electronic technology in general, or what we now call communications technology, marks a new development in human history, a development that is defined precisely in terms of a development in the field of human communication. That development, in turn, is understood as the means to communicate – that is, to speak – over distances previously only ever imaginable. This power of tele-communications technologies to defy distance marks a significant turning-point in history, it is argued, roughly for one or both of two reasons: positively, it promotes a democratic exchange of information within and between communities; negatively, it threatens local communities and regional histories with extinction in the face of mass undifferentiated culture. In other words the technological defeat of *distance* carries the potential to obliterate personal, cultural, historical and political *difference*; depending on circumstances, this potential may be seen as either good or bad.

Secondly, the defeat of distance is achieved by means of separating the physical presence of those engaged in acts of communication from the communicative acts themselves. So, for example, it is of no consequence whether a commercial television channel puts its evening news bulletin to air 'live', in the presence of its newsreader, or whether it televises a pre-recorded version of the news in that time slot, since in neither case is the newsreader actually *present* in the way that two members of a household would be understood to be present in the same room if they were watching the telecast together. The fact that 'presence' turns out *not* to be a necessary condition of human communication (you could have been listening to this chapter on cassette tape, remember, just as you are reading it a million miles away from where I am now) should mean that communication in all its forms, including speech, cannot be grounded in a single authenticating gesture. For if that gesture can be copied (the 'live' presence of a newsreader is indistinguishable from his or her 'simulated' pre-recorded presence, if it is presented as 'live'): if it can be copied, and the copy is indistinguishable from the 'original', then how could it be essential to any test of verification or authenticity? In turn this should mean that the electronic media cannot be held accountable for the breakdown of communication in the family, the decline in standards of conversation, the failure of (post)modern society and so on. They cannot be held accountable because they haven't replaced anything that was *essential* to human communication in the first place, in the so-called natural form of speech; they haven't *replaced* speech, they've copied it. Or rather speech itself always already copies telecommunication, is always already itself a *form* of telecommunication (a point to which we will return in Chapter 9). After all, speech does indeed enable the *extension* of something (whatever that might be: 'the central nervous system', the unconscious, the self); and so, in so far as it extends that something beyond or outside itself, it must be said also to *distance* that something *from* itself. A speaker and his or her words are always distanced from each other, *as a necessary condition of speech*: otherwise the speaker's words could not be reproduced in his or her absence.

Communication is always already, then, telecommunication. But speech is nonetheless understood to be the natural mode of communication against which all other forms, especially those dependent on electronic technology, are understood to be artificial.

If speech cannot be claimed to have priority to or over telecommunications devices, however, then this is because technology as such did not simply occur with the invention of the pen, photography or the fax machine. Technology is not, that is to say, a product only of history. Hence in a certain sense the distinction between communication and *tele*-communications is a naive one, 'for all communication contains within it the notion of the telos, of distance and spacing on the one hand and destination on the other'.[8] To speak is always already therefore to practise a form of telecommunications: the words I speak are always spaced apart from each other, and from me; they are always at a distance from myself and you; and they are always intended to arrive at a destination, in the form of a complete and fully comprehensible meaning, albeit the intention is no guarantee of success.

What this might suggest is that our (postmodern) so-called technological world differs from the so-called pre-technological past in terms only of surface effects. The electronic media might constitute a no more corruptive system of communication than speech, and modern advertising practices in themselves might not (contrary to Blonsky's view as seen in Chapter 2) be responsible for social disorder and moral decay. But it's clear, nonetheless, that technology *is* blamed for the speed at which the world is changing, usually at the level of an increasing gap in social relations between individuals and some ideal of history in the form of deep tradition. What technology brings about, in other words, is *loss* – a sense of lost community, lost history, and therefore ultimately of the lost or alienated self. This 'loss' or 'lack', of course, is seen to be a result of technology's reproductive capacity, its ability to copy 'natural' forms of human communication which depend on the living *presence* of humans (see Chapter 5). Hence the ultimate science fiction horror: that the machines, having learned to copy us, will finally dispense with us altogether.

Nor is this a fear confined to the avant garde. Each of us experiences it, for example, whenever we pick up a telephone and are faced with the prospect of talking to someone who isn't there. As Gary Gumpert argues: '[w]e then become aware of the interposition of communication technology because we are confronted with it. This confrontation is one for which most people are not really prepared.'[9] While it is certainly true, however, that we are not usually prepared for such confrontations at the moment they occur, it is true also that we are trained to accept the same conditions of occurrence as quite natural under other circumstances. I do not go to the cinema, for example, believing that Robert De Niro is actually himself 'there' up on screen, but rather that it's a technologically reproduced image of him; for this reason, it is an in-principle possibility that Robert De Niro could be sitting next to me while I was watching *Raging Bull*. But does the same principle apply to

watching the news on television? Would we not experience a confrontation of a whole other order if in fact the person reading the news on
television were actually sitting next to us while we were watching an
image of them reading the news on screen? We are trained, in other
words, to expect that television news goes to air live, which is probably
its most telling truth effect, so that it cannot (albeit only by convention)
be televised in the absence of a presenter. By the same token, of course, it
would be possible to impose this same convention on films, although
this would certainly have effects at the level of film-making: thus, in
principle, movies could be screened 'live', as they were being made, just
as television drama used to be.

All of which is to ask – why is it only when we don't make contact
with a *person* on the other end of the telephone that we 'become aware of
the interposition of communication technology'? A likely answer to this
question is simply that we *expect* a person to be there, even though their
absence proves their presence to be inessential. In principle, that is to
say, I can just as easily do business with a *computer* over the telephone as
I can with a human being; while (again, in principle) a friend and I could
as easily conduct a conversation through each other's answering
machine as we might do in person.

All the same, most of us would prefer not to have to speak to a
computer and not to carry on conversations with our friends via
telephonic recording instruments. But what might be at stake in this
preference is our fear of recognizing how little it takes for information to
be exchanged, how little is in fact essential to the process. It might be
simply that the telephone has only recently become a source of this fear
due to its changing nature as a policy object, which in turn might reflect
recent technological changes in a global environment of data-transfer
possibilities. Previously, then, the telephone's designated use was
almost exclusively in terms of an 'intrinsic' rather than an 'instrumental'
communications device, a distinction drawn originally by Suzanne
Keller and reproduced in a paper by Ann Moyal. In terms of this
distinction, the telephone's intrinsic communication function depends
on the co-presence of caller and respondent, thus enabling Moyal to
claim that the telephone 'connects families, fosters friendship, supports
the alienated and lonely, and provides a mainstay for a caring society'.[10]
Despite its being technologically possible to achieve, then, it would be
inconceivable, according to this understanding of intrinsic telephone
use, that hotline counselling staff should be replaced by synthesized
voice tapes, for example, or that their professional advice should be pre-
recorded and played to desperate callers in the physical absence of the

counsellors themselves. What must be *intrinsic* to such forms of telephone communication, it would therefore seem, is that they're intrinsically *human*, which is to imply that what they communicate is more than can be quantified as just 'information'; indeed, it is precisely this 'more' which is crucial, and its nature is precisely what cannot be copied technologically, what cannot be reproduced in the absence of human beings. We cannot yet, after all, imagine 'a caring society' that was run by machines – and this in spite of the fact, as everyone knows, that sincerity is easy to fake. No one reading this, for instance, really knows whether I'm sincerely committed to the topic of this chapter, or if I'm sincerely trying to make it 'work', or even whether I'm sincere about the responsibilities of my job in general – though it is probably fair to say that the answer to all these questions is that I appear to be, which is a perfectly sufficient condition for you to believe that I am. As it would be, of course, if I were a telephone hotline counsellor; so why not replace me with a machine?

Used in this intrinsic way (which Moyal, significantly, describes also as 'interpersonal'), the telephone, under normal circumstances, remains 'invisible',[11] having the technological status, as it were, of a very 'low' form of high-tech. Its invisibility, except due to equipment failure or technological error, is precisely a result of the ease with which it can be designated a device of *interpersonal* communication – in the 'human' sense of something *more* than pure information exchange. Thus, by contrast, an increasing designation of telephone use is in terms of 'instrumental' communication which 'binds the business community, exchanges data, pushes commercial and technological activity, and flexes the economic sinews of the nation'.[12] Interestingly, Moyal appears to want to map her use of the distinction between instrumental and interpersonal communication functions of the telephone onto a straight-forward gender division. Briefly: the instrumental function is disembodied, public and 'masculine'; its role is to facilitate new developments in commerce and knowledge. The interpersonal or intrinsic function, however, is very much embodied, private and 'feminine'; its role is to perpetuate community and ritual. The one is therefore pro-ductive, the other re-productive (see Chapter 5). Hence, although it 'binds', the masculine instrumental function also 'exchanges', 'pushes' and 'flexes', while the feminine interpersonal function 'connects', 'fosters' and 'supports'. It is scarcely surprising then that 'masculine' uses of the telephone should be in the service of an abstract 'information flow', whereas 'feminine' uses constitute for Moyal intrinsic service in the preservation of 'a caring society'. It is consistent, moreover, with this gendered reading that interpersonal uses are 'invisible', as so-called women's work is often held to be.

Nevertheless our interest here is in the increasing *instrumentality* of telephonic communication usage. Telecom Australia, for instance, is now offering its 'most valued customers' (among whom I am seemingly included, since I received a Telecom letter addressed to me in such terms only recently) the chance 'to join a new, more personalised service, called NETWORK PLUS'. Operated through a plastic card, the new service promises 'access to improved and expanded Telecom services, special offers and early notification of upcoming developments'. The NETWORK PLUS card itself doubles as a Telecard, 'which is a clever idea from Telecom that allows you to make calls from any phone, anywhere, and have them charged to your home phone account'. Also on offer is another new Telecom service, EASYCALL^{TM}, which can 'turn your ordinary phone into an extremely clever phone for less than 50c a week'. Again the sales emphasis is not on the benefits of high-technology to, say, the warfare industry, but on its service to democracy in the form of increased customer flexibility and consumer choice that user-friendly EASYCALL^{TM} makes possible:

> Easycall is an amazing Telecom service because it can do things that were previously thought to be impossible with your home phone.
>   For instance:
> **Call Waiting** allows you to put one call on hold and answer another ... with the touch of a button.
> **Call Diversion** allows you to redirect calls to wherever you're going.
> **Conference Calls** allow you to have a conversation with more than one person at the same time.

> Other options are available and you can find out more about Easycall and how easy it is to connect by calling the NETWORK PLUS hotline number printed on your card. (Easycall is not yet available in all areas.)

Perhaps it would be instructive here to set this sales pitch against a more critical account of social uses of the telephone, as suggested by Peter White:

> In the short term many new information services will be provided through the basic telephone system. Access to, and use of a standard telephone will be the minimum requirements for access

to many new services. So for individuals and groups which cannot gain access to a basic telephone service in the short term, or where access to a telephone service is too expensive, there is little to be gained by talking about the long-term promises and social benefits of new technologies.[13]

Despite an insistence on the 'personalised' nature of the new NET-WORK PLUS and EASYCALL[TM] systems, then, what is really being sold is customer access to an expanded *instrumental* use of the telephone. This is especially the case with EASYCALL[TM] in terms of its business-oriented service attractions: call waiting, call diversion and conference link facilities are clearly of most advantage to the corporate sector. 'While the future of telecommunications is presented in terms of the benefits which will flow to the corporate sector,' as White puts it, 'there is little public discussion of the implications of these changes for domestic users of the ordinary telephone.'[14]

White is responding to recent policy initiatives that indicate a decreasing government interest in domestic interpersonal telephone use. The emphasis now, as tabled in a Statement by the then Minister for Transport and Communications, Senator the Hon. Gareth Evans QC, published under the title of *Australian Telecommunications Services: A New Framework*, dated 25 May 1988, is on enabling 'all elements of the Australian telecommunications industry (manufacturing services, information provision) to participate effectively in the rapidly growing Australian and world telecommunications markets'.[15] Presumably the shift in policy emphasis towards support of the corporate sector is a measure of the government's belief in the achievement of its main objective: namely, 'to ensure universal access to standard telephone services throughout Australia on an equitable basis and at affordable prices, in recognition of the social importance of these services'.[16] Whatever the justification, certainly the bulk of the 228-page document concentrates on business sector needs at the expense of much consideration of their impact on 'domestic users of the ordinary telephone'. Thus the report concludes that the principal telecommunications objective of the past, to ensure 'provision of telephone services throughout Australia on a non-discriminatory, uniform basis at affordable prices', is no longer a sustainable priority. 'While the traditional objective remains important, it is no longer sufficient, by itself, to meet Australia's future needs for telecommunications services.'[17] In the paragraph that follows this statement, however, *Australia's* 'future needs' collapse into those of the corporate sector: 'The three major factors creating pressures on the telecommunications system are technology, business community needs and the changing world economy.'[18]

Increasingly, as I have been at pains to repeat, policy regulation of the telephone is directed towards instrumental use-designation in the service of business interests. There is little doubt that this will lead to the creation of an underclass of telephone subscribers who won't be able to afford any new domestic services that are made available through corporate spin-off – let alone, of course, the 'telephone poor'[19] who can't afford a phone at all. For the rest of us, however, 'futuristic' phone use will soon become a part of daily life. After all, as Gary Gumpert writes, '[i]t is part of growing up in the twentieth century for the technological to become commonplace.'[20] In time, the 'postmodern' phone will become as 'invisible' as the telephone has been for at least the past thirty years. For again, as Gumpert points out, the telephone has always had the potential to disturb our sense of the real: 'there is *nothing* intrinsic in the medium of the telephone that indicates location', for example.[21] Instead telephony subordinates a sense of place to 'the substance of communication',[22] which occurs not in social space but in 'telephone space', the space of semiosis, as it were, or what Derrida might call the space of writing (see Chapters 3, 5 and 9). This space is made possible, I would argue, precisely because the telephone, like other communications media or 'writing' instruments, does not require the physical presence of users in order for it to be used. It is simply the sending of messages, the transfer of information, that matters – not the arrival of meaning.[23] Or, as Gumpert argues:

> Because site and telephone number do *not* have to be fixed and since the call can be routed to any place the receiver of the call designates, location is subordinate to the substance of communication. As long as the connection is completed, it does not matter whether a call placed in New York City to a specific location in that city is routed to another part of town or perhaps to Los Angeles. If the purpose of the phone connection is to transfer information, if the emphasis is on data and not on personality, location does not matter or, at least, is secondary.[24]

To this it might be added that since this structural possibility adheres in every telephone call, then what Gumpert alludes to as a 'personality' call and what Moyal refers to as 'intrinsic' or 'interpersonal' telephone use, describes in fact a *special* instance of telephony. All telephone communication, in other words, is instrumental – to the extent that even a deviant or interpersonal call can always, in principle, be redirected, be answered by a computer, result in a wrong number and so on. The 'postmodern' phone, then, as becoming possible through NETWORK PLUS and EASYCALL™, is not science fiction coming true. Not even when it develops into the videophone will the telephone of the twenty-first century have become another technological object entirely from that developed by Alexander Graham Bell and others over a century ago. Instead, and this is a deceptively simple point to end on here, the telephone today and of the future is a product of its redefinition as a policy object. Policy-makers cannot change the nature of human communication; they can change only the political organization of communities. Hence we should remain cautious in our assessment of recent policy initiatives in the history of social uses of the telephone.

5. *Situating Technologies: Radio Activity and the Nuclear Question*

I forgot to remember to forget.

Elvis Presley

On a postcard dated 7 September 1977, Derrida writes of an 'old dream' – but one whose 'rarefaction is unbearable for me in writing – the old dream of the complete electro-cardio-encephalo-LOGO-icono-cinemato-bio-gram'.[1] In dreams, at any rate since Freud (if not after Orbison, or Lynch), strange associations occur, though these are never less than always 'postal', always being destined to arrive but not necessarily (as we saw in the previous chapter) at an intended or 'specific location'.[2] Strange indeed, however, is the seemingly unmotivated chain of associations linking Derrida's dream of an arrival which would be 'complete'. Stranger still, the dream is very old – and Derrida is not alone in having dreamt it.

Plato also had this dream, the recurring dream of Western metaphysics ever since. Many kinds of dreamwork could be done on it, but for the moment let us consider the terms of Derrida's analysis above. What kind of writing is dreamt of here? In a word, a writing which is *complete* – in the sense of transmitting heart-felt, brain-driven, SPEECH-centred images in living motion. Nothing could escape this writing, which technologically could not fail to reproduce 'life' as a perfect adequation of itself and thus be completely indistinguishable from its representations in 'writing'. But such a writing remains a dream. Writing, as we know and practise it, then, must always be *in*complete – something must always escape a technology that is always less than perfectly able to adequate whatever it is a writing 'of'.

The extent to which writing can be understood as an instance of technology in general, at least in Derrida's terms, is measured by the degree to which writing and technology are always positioned within logocentrism as lapses or a fall away from nature.[3] In other words, it is nature's *priority* both to and over 'art-ifice' (*physis* versus *techne*) which accounts for writing's status as a lapsed or incomplete form of speech: writing, simply, as *written-down* speech. But what this opposition occludes, for Derrida, is the impossibility of nature's 'own' completeness, as if it were a domain locatable outside writing and wholly separate from effects of signifying practices used to constitute and privilege its status as origin (and here we might recall the discussion of Vico in Chapter 1). Similarly, speech's privilege as the 'natural' mode of communication hides a dependence on writing effects which are necessary to its 'self'-constitution as the origin of truth, from which writing is then able to be seen to have lapsed.

It must follow from Derrida's argument that debates over technology in general will slip into forms of the speech/writing opposition. With few exceptions, this accounts (as we saw in Chapter 3 especially) for technology's fate as a *necessary evil* that threatens the natural order and evolution of human history, human culture and the human spirit. The evilness of this threat lies in technology's power to make us forget our true natures as living social beings: because technology allows us to *do* less ourselves, therefore it is feared that we're *becoming* increasingly less 'ourselves'. By the same token, the *necessity* of technology is its power not simply to improve and prolong life, but to help us remember our true natures *as* social beings by extending the domain of the social to include the whole world. Since this potentially positive effect contains the dangerous threat of 'self'-extinction, however, by means of the species' total memory loss, technology requires continual surveillance, constant monitoring and careful management. In short, technology is beneficial only in so far as it can be managed in the service of what might be called forms of 'nature' enhancement, so that the natural might better propagate 'itself'. What is most in need of management, paradoxically, is technology's seductive death drive, its inherent, 'natural' propensity to reproduce 'lifeless' messages in the absence of their living producers, their original interlocutors and the social history of the circumstances of their production. The exemplary instance here is of course photography, a technology responsible for the complete separation of the sign from its embodiment in social practice, its living use as a sign *of*. Hence the photograph, or the photographic reproduction of a painting, threatens the death of truth by means of its technological *iterability*, its capacity to be photo-copied. Any special significance (Benjamin's 'aura') which might be claimed for the original image as a textual representation of living bodies, real historical subjects, is thus destroyed by the photographic image's constitutive reproducibility which, over countless image-generations, dissipates the *presence* of the subject/s and producer/s of the image to an eventual vanishing point.[4]

But as if there were another way. For the seeming specificity of the photographic image as a technological mode of communication presupposes its difference from a norm: natural modes of communication. Natural modes of communication must be supposed, therefore, to have a life of their own, fully embodied with and indissociable from the lives of their users, which reproductive technologies may only ever (dis)simulate. To the extent that all technologies are reproductive, in so far as they enable endless repetition, unlimited (self-)regeneration of the same, technology in general is therefore on the side of writing, loss (lack), and ultimately of death. To the extent that technologies divide their users from themselves, precisely by extending the domain of the cultural beyond 'natural' proxemics, technology in general is therefore 'written'-down nature (contact lenses copy seeing; electronic mail systems copy interpersonal contact; the telephone copies conversation, etc.). It is this inscriptive, supplementary character of technology that constitutes its danger – precisely because of technology's self-reproductive power. Like writing, technology operates in the absence of *specific* agents and, even more frightening, in the absence of *purpose*. No one in particular is required to operate a vacuum cleaner, and neither are that machine's uses and effects confined to a notion of purpose-specificity: it might just as well be used as a sound-effects instrument as a machine for sucking up dust. Hence the need to *manage* technology, to *control* writing. Because, if the example of the vacuum cleaner seems trivial, we need ponder only the possible uses and effects of *nuclear* technology to comprehend the fate that awaits us all: complete sudden death.

Other fates await us too, of course, perhaps like long-lost letters gone astray. We just have to sort our way through all the mess. How we might go about doing so is undeniably urgent and crucial, but it's not as simple as resolving the fate of technology by situating the nuclear question on the side of speech. The logocentric choice is no choice at all: that way the future is already written. Instead, or perhaps by the same token, we need to take account of the 'postal' effects of our thinking on technology, as if the 'technological' future or the 'natural' past were all there is – a choice between techno-doom and self-determination.

In order to rethink technology in terms of the postal, first we must acknowledge the divided nature of the spoken word. In turn, this will entail a recognition of the 'writerliness' of seeing, speaking and all other forms of 'interpersonal contact' such that these may no longer be understood as 'natural' modes of communication or cognition against which the contact lens, email and the telephone stand as technological devices for copying the preconditional faculties of our 'humanness'. For this reason, John Berger was wrong: 'seeing' does *not* come 'before words'.[5] Seeing does not arise from its precondition – sight – any more than sugar comes from sugar cane, or butter from cows. Yet it's interesting that Berger's proposition can make sense only within ways of thinking that habitually take for granted the priority of speech to writing: this is the model that allows there to be a 'natural' language, which Berger calls 'seeing', having priority over 'artificial' languages (simply 'words', as they are spoken or written down, for Berger) whose protocols of use have to be learned and practised before being applied – hence they don't come 'naturally'.

Yet again it would seem that Berger's is 'the old dream of the complete electro-cardio-encephalo-LOGO-icono-cinemato-bio-gram', though doubtless he wouldn't see it that way. Once it is possible to do so, however, and the virtuality of the dream is endlessly deferred from becoming actual, the 'postal' nature of the spoken word comes into play. This would be to acknowledge that words as well as letters often go astray; that missives and missiles can be interfered with. No less than writing, then, speech is subject to unavoidable effects of misdirection, misattribution, miscalculation and so on, which together form a set of mailing mishaps that the postal system is designed to overcome, but never completely. There is nothing to prevent a letter being misaddressed, nor to prevent even its non-arrival. Neither can the medium of speech guarantee to deliver the intended meanings of speakers, any more than a postal system could function without a dead-letter office.

In this always divided sense of the spoken word, writing can no longer be understood as a technology of differencing, distancing and delaying effects – at least not 'writing' in the old sense of a technology for copying speech. Hence if there is no 'natural' mode of communication as such, there is strictly speaking no technology that has a power to distance or divide us 'from' ourselves. Equally, there is no technology that can join us 'to' ourselves, despite the offence this claim must give to much rhetoric surrounding the technology of radio.

The asserted power of radio to perform what I have maintained a technology to be incapable of performing (repair work at the level of the sign, as it were, or on the order of subjectivity) goes back at least to Brecht, who wrote in 1932:

In our society one can invent and perfect discoveries that still have to conquer their market and justify their existence; in other words discoveries that have not been called for. Thus there was a moment when technology was advanced enough to produce the radio and society was not yet advanced enough to accept it. The radio was then in its first phase of being a substitute: a substitute for theatre, opera, concerts, lectures, café music, local newspapers and so forth. This was the patient's period of halcyon youth. I am not sure if it is finished yet, but if so then this stripling who needed no certificate of competence to be born will have to start looking retrospectively for an object in life. Just as a man will begin asking at a certain age, when his first innocence has been lost, what he is supposed to be doing in the world.

As for the radio's object, I don't think it can consist merely in prettifying public life. Nor is radio in my view an adequate means of bringing back cosiness to the home and making family life bearable again. But quite apart from the dubiousness of its function, radio is one-sided when it should be two-. It is purely an apparatus for distribution, for mere sharing out. So here is a positive suggestion: change this apparatus over from distribution to communication. The radio would be the finest possible communication apparatus in public life, a vast network of pipes. That is to say, it would be if it knew how to receive as well as to transmit, how to let the listener speak as well as hear, how to bring him into a relationship instead of isolating him. On this principle the radio should step out of the supply business and organize its listeners as suppliers. Any attempt by the radio to give a truly public character to public occasions is a step in the right direction.[6]

Brecht could hardly be accused of techno-paranoia here, and therefore might be said to be out of temper with the times. But we should not miss understanding that his projected charter for radio's proper use is dependent on the speech/writing opposition for its insistence on the *dialogic* nature of the broadcast medium (precisely the affirmation invoked today on behalf of the internet).

Although more concerned to propose a politics of radio than to develop a theory of radio as technology, such a theory nonetheless underlies Brecht's politics in the form of his commitment to radio's speech potential and thus to its politicization as a socially *interactive* medium. If properly controlled, in other words, radio's speech-like capacity has potential to extend the domain of interpersonal contact among listeners (who would then, in a sense, be able to 'write', to practise, to engage with radio instead of passively receiving it), rather than consisting merely of a capacity for 'prettifying public life'. The purpose is to find therefore a way of actualizing, activating radio's potential 'as' speech, which would be to overcome its present use 'as' writing – a use that constrains radio to re-produce only the surface effects of community living when it could be producing a deep-rooted sense of real human history, of community – of culture. The urgent need of such a purpose, of course, is due to radio's power *as* a technology to make us *forget* our history – the danger being that we risk remembering only what is broadcast *to* us, what is written down and therefore distanced *from* us, as the truth.

Similarly, for Brecht (and here he shows himself to be not completely out of joint with his times), the dangerous writing effects of an improper use of radio extend also to the misuse of literature. 'We have a literature,' he writes, 'without consequences, which not only sets out to lead nowhere, but does all it can to neutralize its readers by depicting each object and situation stripped of the consequences to which they lead.'[7] Once more the problem is a lack of interaction between a technology and its users, who need to be 'actualized' if the technology is to be approved for its exchange potential, its two-sided, dialogic effects. In the present case, that approval can be given only if the technology is used for its potential to produce what might be called speech-like writing, or (good) literature. The same technology, however, also produces a literature 'without consequences' that fails to make contact with its readers, threatening to replace their living memorial experience with a 'prettifying' supplement to their lives.

It would thus seem from this example that the speech/writing opposition, when it controls the forms of debate over technology in general or radio in particular, determines the positivity of its object in terms only of the dialogic. 'Bad' radio, as defined by Brecht, is straightforwardly *monologic* (there is no suggestion of a conspiracy, albeit this is just a slip away, but simply the assertion of radio's *misuse*): it speaks to or for an audience, whose only value to the broadcast institution is their status as the *silent* majority – voiceless, inactive, collective. In short 'bad' radio denies its listeners a *speaking* position, thus determining their subjectivity as mute and passive. 'Good' radio therefore would give 'voice' to its listeners, construct its audience as 'speaking' subjects and provide for interactive dialogue between itself and different audience communities.

An ideal form of such radio, a perfect social use of broadcast technology, would thus appear to be talk-back. But that talk-back hosts and their producers wield unequal power – in any exchange with callers, any act of 'communication' with a listener – should be sufficient grounds for hesitation in claiming radio as a means to greater democracy, founded on a disseminating power to spread the spoken word. Nor is this to recognize only the limits of the broadcast *institution*, which might be overcome (as they would for Brecht) if radio audiences were permitted a more active role in the means of production. The call for radio to fulfil its potential speech-equivalence, in other words, forgets the equivalence of all technologies to writing. Such acts of forgetting are dependent on the priority of a 'natural' mode of human interaction and self-cognition (speech) to all other forms of human exchange (writing), which are thus defined as 'artificial' and feared for their potential to erase our memories. Before any technology can be discussed in terms of its speech-equivalence, we must forget to remember that it *is* a technology.

Adjusting the Frequency

New technologies have a power to excite, or to arouse fear, as things in themselves. The computer and software I'm using to write these words no longer seem any more 'technological' to me than a ballpoint pen. It is the words that I can produce with these technologies of which I am conscious as 'things', not necessarily 'in themselves' but at least (among other senses) in the forms of their materiality.

I have evolved into a good cyborg, fully embodied as one with 'my' machine. It is only rarely (when the machine malfunctions, or there's a power failure) that I will curse the computer, as I would curse the pen for running out of ink on me in mid-sentence; curse it like a dog. Otherwise when I express impatience with it, it's as if I'm talking to myself. Such is the fate of the good machine: to facilitate our evolution into cyborg beings – not simply our adaptation to technological 'literacy' but our constitution as machinebody selves.

Of course, this process has a history and it is perhaps only with the invention of new technologies that we become aware of it. At present we are thrilled or terrified by virtual reality, for example, because the newness of this technology underscores its status as (a) thing: technology (in) itself. We cannot yet 'see' but only imagine past that thing to what it might allow us to make possible, let alone its effects. No one reads a book and sees a printing press, but it would be difficult to overlook 'technology' in discussions of the coming virtual real.

Similarly, we do not see technology when we listen to the radio. This has not always been so, but nowadays radio gives us nothing to 'see'. In the sense in which I am using this word, however, it is precisely the fate of technology to become *unseeable*, so that we don't see television or cinema either. We don't see technology at all; no more than we see a pen or a washing machine, except as a tool-object which enables us to 'do' something with it, and quite unlike the sense in which we 'see' aesthetic-objects (whether of the visual or the nonvisual arts: paintings as well as poems, in other words). To *see* a thing is to value it *as* a thing, to see it as a thing in itself – with self-integrity and a constitution of its own. Like a painting, a poem, or a person. By the same token it is *not to see all of that thing*, precisely because its integrity *is* its own and cannot be captured completely by sight; something must always escape from view.

In these terms, according to this metaphor, we might say that we are oblivious to technology and that the things we see are always at the same time and paradoxically mysterious to us. This would make us blind to technology and describe what we see as 'art'.[8] It would mean that we do not fully integrate with the things we see and that technology is 'invisible' because we're too close to it (or vice versa) and so it seldom falls within our field of vision, within scope of our being able to relate to it as or by other than practical means – means acquired like writing skills to instantiate our presence in the social or our being in the world. In other words, to see a thing is to be conscious of its *distance* from us (or vice versa); it is to be conscious of the thing which is outside, other. But we incorporate *into* ourselves such things as the uses and applications of tools, instruments and technologies – things that we see only for their uses and therefore don't see as things (unless, of course, we don't know how to use them). To become skilled in a technology is thus to consume it and afterwards consign it from sight as a thing in itself, its only value as a thing being always already whatever it allows us to do. Whatever that is, it is afterwards all that we see: clean clothes, a polished floor, a page of writing.

On the other hand, perhaps, by our corporeal incorporation of technologies we are living out our species' fate. It is not that we have become cyborgs only recently, but that since Descartes we have until only recently been able to rethink (that is, to recall) the subject's self-constitution in the world in terms other than a reliance upon its 'own' consciousness and determining powers of self-reflection. One of the first moves in this reconceptualization of the subject/object split (a turn, as it were, which might also be a return) is Heidegger's understanding of the priority of our being in the world to the constitution of our being as reflecting subjects. Before it can become conscious of the self-identity of its own being (t)here, *Dasein* (the subject in existence) first of all has to be grounded in an orientation to the world in general, the world of object-others that exist in space as well as time:

> The possibility of indicating is grounded in the constitution of orientation. Indicating lets a 'there' be seen and experienced. This 'there' brings with it the discovery of the corresponding 'here' of indicating and of the indicator. The fact that environmental signs are encountered, understood, and used means that being-in-the-world ... is as such oriented [and] because oriented Dasein is corporeal Dasein, corporeality is necessarily oriented. The orientation of apprehension and *looking* articulates the 'straight ahead' and the 'to the right and left.' Dasein is oriented as corporeal, as corporeal it is in each instance its right and left, and that is why the parts of the body are also right and left parts.[9]

It is by means of positing the self's pre-reflexive faculty or condition of orientation to the world that Heidegger's critique of the Cartesian subject's self-constituting powers of reflection breaks with Enlightenment wisdom, simply by positing the priority of that condition *to* our determination as 'selves'. Selfhood, then, is always already problematically imbued with otherness; difference grounds identity as a precondition. *Dasein*'s corporeal orientation thus predisposes it to constitute itself as a seeing subject, albeit with a bad memory: for it is easy for *Dasein* to forget its own *situatedness* within the world, an act of forgetting which is necessary to the constitution of *the world outside itself*.

I do not seek to overstate the 'radical' effects of this Heideggerian breach with the Cartesian (that is, modern) tradition of the unified subject. After all, as Derrida has shown, Heidegger's move does not escape the politico-philosophical system it threatens to undo, despite reaching further back than anyone into the history of Being for the ground, the origin, the essence of human consciousness: it still remains a philosophy indebted to a notion of telos, even if that telos cannot easily be described within the limits of traditional semantics and thought.[10] But Heidegger at least allows for a putting into question of the whole (thing), whose integrity depends on a classical division of subject and object and which Derrida has since subjected to greater effects of undecidability in the form of his critique of the Western epistemic tradition's reliance on the binary sign as the condition of the appearance of truth.[11] For present purposes, then, Heidegger's shifting and stretching of the philosophical limits of the grounds of subjectivity may be claimed as an opening move in the formation of a postmodern piety – the decentred subject, who is divided from itself and therefore lacks a critical distance from the world in which to speak other than partial, situated 'truths'.

But if Heidegger is the postmodern subject's surrogate father, Derrida is that subject's Dr Frankenstein. Once the classical sign was seen to contain or conceal an excluded middle, a third term – signifier (slash) signified – that middle became a gap, a present absence, a space of difference in which relations of connectedness between signifiers and signifieds no longer have to operate on a horizontal plane (as if the slash were always only a lever) but rather within a kind of phase space where holding patterns of connection can be described in terms only of strange attractors, if at all.[12] According to classical (Pythagorean) geometry, the classical (Saussurean) sign looks like this:

$$\text{Sr} \;\Big/\; \text{Sd} \;=\; \text{Sr} \;\text{——}\; \text{Sd}$$

But by rethinking the sign in terms of nonlinear relations of contiguity or connectedness within, we may find that its topology looks quite different. Instead of a single uniform line occupying a flat surface, we might discover that the sign's internal system of relations can be traced to form a set of warped trajectories oscillating around a strange attractor. Hence it might look something like this:

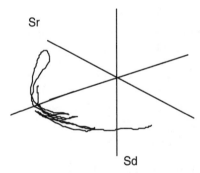

If the first figure describes a state, the second describes a process. Thus we need a semiotics of dynamical systems, a science of signs which is a science of process rather than of state.

Such a project is already under way, of course. As long ago now as 1966, however, Derrida anticipated at a conference on the human sciences at the Johns Hopkins University, a moment that might be claimed retrospectively to mark the emergence of post-structuralism within the English-speaking academy, that any transition from a science of state, a philosophy of the centre, the ground, the essence, towards a science of process would be received with horror. Any movement away from the notion of centred structure could be understood only as a denial of humanism, and that way lay not science but science fiction, the fantastic, *the gothic*. When there is no longer a choice between Rousseauistic nostalgia for a lost origin and Nietzschean affirmation of signplay, then

> there is a kind of question, let us call it historical, whose *conception, formation, gestation,* and *labor* we are only catching a glimpse of today. I employ these words, I admit, with a glance toward the operations of childbearing but also with a glance toward those who, in a society from which I do not exclude myself, turn their eyes away when faced by the as yet unnamable which is proclaiming itself and which can do so, as is necessary whenever a birth is in the offing, only under the species of the nonspecies, in the formless, mute, infant, and terrifying form of monstrosity.[13]

What Derrida did not predict was that, in time, this brainchild would be given a name – deconstruction – and, although it would continue to be regarded in some circles as the 'Rosemary's baby' of philosophy, it would also give rise to a cult of pious disciples. Even so, as Derrida put it back in 1966,

> although these two interpretations must acknowledge and accentuate their difference and define their irreducibility, I do not believe that today there is any question of *choosing* – in the first place because here we are in a region (let us say, provisionally, a region of historicity) where the category of choice seems particularly trivial; and in the second, because we must first try to conceive of the common ground, and the *différance* [or differance] of this irreducible difference.[14]

Yet still we continue to choose: between philosophy and politics, speech and writing, technology and art. But that we *must* continue to do so, under certain circumstances and in certain situations, is simply to underscore the triviality of the *question* of choosing – sometimes there is never any choice but to act against. Sometimes we need a politics of state, not a philosophy of process. At these moments we must act to *close* the gap within the sign, or *as if* the sign already looked Pythagorean rather than chaotic. At other times, it is important to recognize 'the common ground' of opposing views, as well as their differences, so that neither extreme can stake a claim to resting on the grounds of truth.

This is not always easy. Debates over technology, for example, seldom appear to allow for any slippage between, let alone within, the two sides. Either technology is valorized for its social benefits (and so we tolerate vivisection, nuclear proliferation, environmental damage, etc.) or attacked for the social, environmental and sometimes spiritual costs those 'benefits' incur. In so far as each of these positions is teleological, one might argue that both the hippy and the corporate industrialist share a common philosophy. By the same token, this should not be to suppose that all anti-technologists are 'hippies' and all pro-technology advocates are capitalist tycoons, despite the habit of each side to construct the other in such terms. But what it suggests is that each side *requires* the other in order to construct and maintain its different 'self'-identities, and in this respect the forms of those constructions share a logic, a way of thinking, which favours a division between an inside and an outside of its 'own' ideological allegiance: progress vs. preservation, culture vs. nature, Homo Faber vs. Homo Spiritualis – and endless other configurations of the speech/writing opposition.[15]

As Zoë Sofia points out, however, apropos the apparent contradiction within US conservative politics' drive for increased nuclear armament and its support of the anti-abortion lobby (a politics which, on the one hand, flirts with species death while stubbornly refusing, on the other, to sanction a single death), the force of this politics' underlying philosophy actually combines these seemingly opposed positions into a single, unified position. In other words, the inside (the ideology of progress, let us say, in this instance) can be shown to contain the outside (the ideology of preservation) within itself; it can be shown to contain its opposite and still appear 'whole'. How it manages to do so, Sofia argues, rests on a certain trope of gestation, a birth metaphor, which she maintains is masculinist through and through. In short, we are always already we; our conception is our birth, our moment of coming into subjectivity. Pro-Life rhetoric thus collapses the adult into the embryo, with the result that

> The embryo faces no alternative futures, but one single destiny, which is moreover collapsed back onto all previous states of being, allowing the conceptus to be spoken of as a 'tiny person' and the deliberate arrest of its development equated with homicide. Contrasting with this *collapsed future* tense of anti-abortion rhetoric is the *future conditional* of feminists, who understand conception as an

occurrence with a number of possible outcomes, to be determined by the future events or decisions which might influence or terminate its development.[16]

Who we are, then, seems to bear no relation to our corporeal beginnings but rather to a male reproductive fantasy of species regeneration 'without the aid of woman'.[17] Tropologically, this is how the collapsed future writes itself and gathers in pro-technology and anti-abortion rhetorics to form a single discourse:

> The collapsed future tense lies at the heart of our culture of space and time travel. It is the 'bound to be' of the ideology of progress, operative in the discourse of those who tell us that since nuclear reactors, deep-sea mining, Star Wars, and space colonies are inevitable parts of our future, we might as well quit griping about their bad side-effects and get on with making the future happen; after all, there's no time like the present. Trouble is, the collapse of the future leaves the present with no time, and we live with the sense of the pre-apocalyptic moment, the inevitability of everything happening at once.[18]

To restrict the scope of this discourse to view human history, human subjectivity and a global future from a privileged site, Sofia argues the imperative of stressing 'the essential qualities of female embodiment' over the (ungendered) volitional subject in discussions of our natures and the destinies they empower.[19] At least she does so with an eye to feminist control of the abortion debate, so that the grounds might be shifted from under the presently controlling masculinist ideology of progress which in turn both rests upon and promulgates a patently fantastic theory of reproduction as auto-generation. That such a shift would instantiate new grounds for making decisions, however, is perhaps the kind of political necessity that makes choosing trivial.

It is not a question of debating *the lack of choice* which Sofia's position demands, or of this demand's antinomous relation to her stance against the '*collapsed future* tense of anti-abortion rhetoric'. Certainly it is to see, as Derrida remarked in 1966, that binary opposites are never mutually exclusive, that there is always 'common ground' as well as 'irreducible difference' between them. But philosophical undecidability, as it were, can never be an excuse for political indecision. We must always remember to choose, as Nietzsche believed, and to forget those choices.

Switching Channels

Rememberingandforgetting.[20] By such a double movement (a unit-without-unity) it is perhaps possible to advance beyond the present deadlock, the *détente* of politico-philosophical positions with respect to questions of technology that always, it seems, invoke the nuclear question. If this is so, then the available positions to be taken up with respect to radio should serve as an instance of the nuclear debate in general. These positions, themselves caught up in discourses on democracy, power and information, split roughly into seeing radio either as a medium for conveying the truth by means of spreading the word or as a tool of any dominant culture's investment in displacing real politics onto illusions of citizenry, community and consensus. Radio is seen as speech *or* writing, in short, which is how technology is viewed in general. What is interesting about radio in particular, though, is its status as a very low form of high-tech, based on its proximity to speech.

Even opponents of radio are not opposed to radio as such, but rather to its control by the 'wrong' interests. In this sense there *are* no opponents of radio – no available position to adopt, as there is with nuclear technology, against radio in itself. Even to argue against radio's 'writing' effects, its power to create illusions of democracy, say, would not be to preclude the possibility of radio being used in the service of more authentic 'speech' effects, if the voices of the people, perhaps, were given greater air time or the people themselves put in control of the means of production. 'Access to the average,' as Derrida writes, 'is often [though the qualification is surely important] a form of progress.'[21] It is to this extent that radio, unlike television, is lent a force of 'intimacy' precisely as a medium in itself, with a power to effect interpersonal ties among listeners and production staff which technologically is closer to the telephone (see Chapter 4) than television.

This would seem to distance radio from the nuclear question, for the very reason of its power to forge and strengthen communities without threatening their extinction. Whatever radio's 'bad' side effects might be, in Marxist terms these are superstructural rather than basic: they belong to the order of ideology, not fundamental history. Since television, at any rate, radio's position as a mass communication medium has been weakened, de-glamorized, forgotten; we no longer 'see' it because we've become oriented to its being in the world – it can never be mistaken for a monster in the offing. Therefore we can debate its effects less urgently – less urgently than those of television whose higher-tech status lends it a degree of appearance to virtual reality and places it more clearly on the side of writing, and those of nuclear technology which threatens to blow us up.

But this has not always been the case. Many examples could be chosen to show the uncertainty principle at work in the development of a social orientation to the new technology of radio. Like other new technologies, it was not immediately obvious what radio was 'for' in its early days. Concentrating on the formation of a notion of 'light entertainment', for instance, Simon Frith argues the deliberate policy-making of the BBC in the 1920s on behalf of a broader cultural agenda directed at edifying an audience of fully volitional, 'English' subjects. BBC programming, at that time under the directorship of John Reith, was thus driven by a politics of 'public service' which sought to address the public in terms of 'nation' and conceived of service in terms of inscribing that public into the self-identity of its own 'Englishness'. While the hidden agenda, as it were, was to preserve a notion of English culture, the most efficient means of achieving it was to displace that agenda onto a paternal protection programme of defending the public's individual subjectivities from the twin threats of massification and Americanization, which were really the same monster (still rearing its ugly heads, as we saw in Chapter 2). Hence a BBC-commissioned report in 1929 on the 'ramifications of the Transatlantic octopus', or the increasing acceptance of American popular culture on British radio and in British music halls and cinema.[22] In contrast to the *speed* of this encompassing spread, which the broadcast medium was at that time best suited technologically to facilitate, the BBC actively sought to slow radio down: moments of silence were broadcast between programmes, to remind listeners not to forget that they were listening, and public statements appeared in the specialist press on how to maximize one's listening pleasures by improving the art of one's listening skills. In short the Reithian regime taught the listening audience *how* to listen – by constantly reminding it not to *forget* that it was listening, which was always to engage in an act of choice, so that listening became a conscious activity performed by an audience of conversant, not complacent, individuals.

Under Reith, then, the BBC solved the hazardous writing potential of the new technology to make people forget their subjectivities (or identities), by deferring solely to radio's curative speech-enhancing properties as a medium of remembrance. In this role, Reith weighed up the paradox of the new technology, as did King Thamus when presented with the gift of writing, but came down on the side of its power to help men and women recall their history and place within a national culture.[23] Perhaps, had he been given Thamus's choice, he too may have decided to reject the new technology, though Reith was obliged to accept the fact of radio's being (t)here and to come to terms instead with a means of eradicating its poisonous effects from its potential to perform as cure. Thamus was luckier, on the other hand. He could reject the gift of writing from the god Theuth, offered on behalf of all men, for what he reasoned must be the poisonous effect of lifeless, disembodied inscriptions on the power of living, memorial truth. In time, men would forget their memories, having grown used to recording them at a distance through means of the new technology. While writing thus seemed to offer a cure against forgetfulness, and so to free history from the haphazards (the chance encounters, the adestinal effects) of human remembrance, Thamus reasoned that in fact the 'truth' which writing promised to record was at best only a dangerous supplement to the 'real' truth embodied in the speech of the mind (see Chapter 3). Their memories having been made inactive by writing, men thereafter would be at the mercy of the new technology and its effects – effects which would become increasingly difficult to locate as the new technology became decreasingly 'new' and visible over time, and which would make it impossible to sort out the truth from among all the lies and slippages.

John Reith's valorization of radio *activity* as the only proper justification of broadcast technology, and the only means of overcoming the radioactive writing effects of the nuclear question posed by the threatened apocalypse of British culture in the face of escalating 'Americanization', might therefore in these terms be understood as Platonic. It must also, to the extent of being profoundly logocentric and if Derrida is right, be on the side of Pro-Life and Pro-Armament rhetorics. To this extent, moreover, the question of the defence of a national culture, a putative 'whole way of life' which is at the same time regionally specific, is itself no less caught up in the dilemma of writing, the nuclear question in general. For when it is a question of survival or extinction, the question of choosing becomes trivial; it can no longer be a matter of 'choice' that we must choose between the bio-politics of being human and the techno-politics of militarist propaganda. But nor should we suppose that our decision can escape the dilemma of writing, or that it can *solve* the nuclear question.

All writing is apocalyptic, in the sense that it is always threatening to arrive, like The Final Solution or the Last Judgement, or *The Day After*; it is always in the act of answering the nuclear question. But that question remains, at present, undecidable (despite recent historical events), as it must always do. *Détente* will never inch towards nor even surpass, but will suddenly *have been* surpassed by *denouement*, the day after. When no one is around to remember. Meanwhile, we must all suffer the after-shock, the radioaction, before the fall-out occurs. We must all live with the nuclear question, as we have always done, never forgetting that this time the threat is real. Species extinction, environmental annihilation, planetary obliteration: these are our chosen trivialities. We approach them, as we have always done, with our fear of the unknown, our anxiety in the face of a dilemma, and seek refuge in what we know: our faith in the vitality of life, the corporeality of living.

Yet we must face them as cyborgs, fully adapted to our adequation by machines, accustomed to our constitution in writing. This is our fate, as it is the fate of technology in general, and we must forget to remember to forget it. For the effects of after-shock, the space of radio activity, cannot be controlled by either side: the airwaves are full of interference, full of gaps. On both sides of the channel there is static, and always the possibility of receiving a signal from afar, the chance of overhearing other voices – ghostly voices, even – from the other side.

Nor is this to feign solemnity, or at least not only for its own sake or that of bathos. For in moving beyond semiotics we continue to open ourselves to unexpected ways of relating to things in the world around us, and to do so without looking for a systematic approach or a consistent method. 'Cultural criticism,' writes Darren Tofts, 'is not a pure art. It has more in common, in fact, with T. S. Eliot's industrial portrait of the poetic imagination at work, in which the incongruous and the ill-fitting, such as Spinoza, falling in love, typing and the smell of cabbage cooking, are drawn together into new syntheses under the force of invention.'[24] Now of course while this is not to say that 'anything goes', it certainly does help to affirm the necessarily inventive nature of our project, if not of any project of cultural criticism in general (see Chapter 9). This is neither to glorify invention nor to 'disciplinize' what we are doing here by calling it a form of bricolage. After all, as Derrida has famously pointed out in the essay 'Structure, Sign and Play in the Discourse of the Human Sciences', bricolage – as a mode of cultural criticism that privileges the readiness to hand of whatever happens to be on offer – is nothing terribly new; indeed it's not even exclusive to cultural criticism particularly or to humanities discourse generally, since every scientist and mathematician worthy of the name is a bricoleur of sorts.[25] Yet there can be no doubt that while it is possible to overstate the importance of invention to cultural criticism, some forms of cultural criticism (Saussurean semiotics, for instance) have sought to suppress the aleatory and asystematic relations between things in favour of imposing an order on them by way of a regulating method. We shouldn't forget that Saussure inherited a 'discipline' without any principles (see Chapter 2), but we need to remember that his over-correction of the problem put semiotics in debt to an all too conservative faith in the metaphysical certainty of things. In continuing to try to move beyond semiotics, then, we need to continue to try to find ways of causing trouble not for semiotics, but for metaphysics (remembering all the while that to cause trouble is not necessarily to surpass or bypass). Since the logocentrism of metaphysics is ultimately a phonocentrism (the sound of the human voice as the touchstone of all truth), let's turn now to a consideration of the properties of sound in such a way that we might hope to do a little mischief.

6. *The Sound of a Dream*

Sounds are never something we simply just hear; we never just listen to sounds. We don't just listen, because there are no sonic objects for listening to. There are only sonic events.

That sounds can be transmitted across space and recorded over time is no proof at all of an object-value for sound. Like the objectifiable written word, the sound-object is a theoretical convenience: the one is always subject to event-ualities of being read, the other to being heard. In this way it may be possible to hear the sound of one hand clapping, on the model of narrating the event of a dream.

Dreams aren't nothing, on say the model of a square triangle, but what they are has no materiality beyond so many electrical impulses in the brain. Yet dreams are not reducible to these impulses, any more than sounds can be reduced to precise configurations of pitch, tone, frequency and other variables. No less than dreams, sounds are irreducible to what they are made of. But what each is made of (neurological and sonic impulses) lends them a certain object-status in so far as these impulses are able to be recorded, even though it would be absurd to suppose that an EEG read-out of a dream was actually a record of the dream itself.

Imagine if I took along one of my dreams in the form of a neurological chart to a psychoanalyst and expected him or her to read it for symbolic content, as if the shape of this peak meant that I wanted to sleep with my mother and the shape of that trough that I wanted to kill my father. Dreamwork isn't palmistry, in other words. But still it is worth noting that, for psychoanalysis, the form of the dream that is of least professional interest is in fact the form which is most objectifiable. Every other form will do, including a lie. So dreamwork is in principle able to be done on a drawing, a photograph or a theatrical performance of a dream, but not on a chart of neurological activity, and it would not matter to the analytic process (although of course it would matter to the analysis) whether any of these forms were deliberately misleading.

Contrary to an earlier remark, then, it would seem that dreams are in fact *not* dissimilar from square triangles, except that the latter can have absolutely no material form whatsoever. A square triangle can exist only in the form of a desire for it to exist, and this is of a slightly different order of being from that of a dream. While dreams might *express* desires, it is precisely in this sense that the dream acts as a carrier of meaning and can to such extent be understood as semiotic. One might therefore be tempted to say that dreaming is a cultural achievement; that dreams are no more 'of' the natural world than butter or plastic.

But to suppose from this that somebody could choose to stop dreaming would be as preposterous as thinking that someone who chose to live on a desert island had escaped culture. The choice, in both cases, is simply unavailable. This is not to deny that dreaming, while it is clearly at some level involuntary, is perhaps still a cultural practice, no less than writing or kissing (which might also be involuntary).

Unlike writing, though, dreams (so far as we yet know) do not leave behind traces of themselves, except in a form of degree zero semiosis that is effectively unreadable. To the extent that dreams are unrecord-able, but nevertheless still 'real', they are more like events than texts (at least in so far as 'text' describes a cultural object that requires to be read or interpreted and which can circulate, precisely *as* an object, indepen-dently of its producer). But dreams can of course be turned into such texts, by means of some form of inscription – and it's at this point that the dream-event becomes, like a page torn out of a diary, something to be scrutinized at a distance.

However, a page from a diary is not just any page of writing: it isn't like a legal document, a newspaper column or even a poem. Ideally, diary entries are personal and spontaneous; their ideal genre is that they don't have one, and their ideal reader is the diarist him- or herself. They are in fact like dreams, in other words. Precisely that: *in other words*.

It is only in the form of other words (other pictures, etc.), as *text*, that dreams are able to be recorded. These other words, ideally, are a transparent textual form of a memory-event. Once again it is important to note that such texts cannot be produced from a neurological chart (though other texts can be), which cannot act as a memorial substitute. Memory can be supplemented only by writing. Whatever memory is, in other words, it can be known only via forms of supplementation, forms of writing, whether verbal or inscriptive. And to the extent that memory is writing dependent, we can say that memory *is* writing (and writing, memory) in so far as it is never available in some pure, whole and original form outside of writing.[1] Nor is there an outside-writing form of the dream, a square triangle or the sound of one hand clapping.

In so far as this is true for remembering, I think it's true also for listening, despite the fact that sound is recordable and therefore, in its textual form, plastic, able to be cut up and played backwards. But plasticity is of the nature of texts, of writing in general, and cannot be confined to sonic texts as a special feature of their aurality (or their orality) based on the object-value ascribed to them by virtue of recording practices and mechanisms. A sound is no more plastic than a memory or a dream, and this is certainly no less true for the object-value lent to sound by being able to be recorded.

Sound's recordability, then, which appears to be the condition of its reversibility and hence its plasticity, is not in my view its most interesting feature. It is not what appears to be the essence of sound that makes sound interesting, in other words. So instead of coming at a theory of sound from the inside, as it were, I prefer (no doubt perversely and cautiously at the same time) to approach the matter from an outside, in the form of dreams.

Dreams are as insubstantial as square triangles, while also being different. If dreams never quite just 'are', at least they can be said to happen; but not even this can be said of a square triangle. And I think the same is true of sounds as it is of dreams: sounds and dreams are event-like, by which I mean *they happen*, but without ever coming into being as objects. Sounds, like dreams, are never capturable by technology; so that any sound recording is in fact of the same ontological status as a photograph of a ghost, which could be likened to a photograph of a square triangle or indeed of a dream.

And so we arrive at a continuum – of sounds, dreams, square triangles and ghosts. Strangely, of these it would seem that it is only sounds that must always remain unable to be photographed – if it were to be believed that ghosts are real; that somehow a square triangle could exist outside a desire for it to exist or beyond the present limits of human cognition (in which case it might exist already, unbeknown to anyone); or that in the future it will be possible to photograph our dreams. Each of these 'events' is at least able to be imagined: I have a sense, for example, of what a photograph of one of my dreams would look like, etc. But the idea of photographing a sound seems to make no sense at all.

This is not simply because sounds cannot be seen in order to be photographed, since I do not believe that ghosts, dreams or square triangles can be seen either. But they can be imagined in this way, and that's the crux.

So what if we began to think of aurality in terms of visuality: what might this produce? What would sounds 'look' like as a consequence of such thinking? By the same token, what would the visual 'sound' like if we began to think of the specular in terms of the sonic?

I am not asking these questions in order to inspire new forms of avant-garde practice in relation to sight and sound. Instead, what I am suggesting is that sounds are heard through the body, that they resonate corporeally and not only on the tympan. Certainly I think this is true of the sounds that I make myself, especially through my voice. My voice, I think, no less than my imagination, while it is able to be prosthetized, exists in such relation with my body that prevents it from being an effect of the operations of some part of my body alone. This is why a recording of my voice, coming to me from outside my body, always has the quality of a prosthetic device; just as, were I to write one, a novel might seem a kind of prosthetics of my imagination.

For this reason, then, I prefer to think of a corporeality rather than a pure aurality of sound, in so far as this allows for sounds to be seen and felt, as well as simply heard. One may even go so far as to postulate an olfactory of sound. In such a way sound is able to be thought in terms of a different set of codifications, which does not rely on the relation of a hearing subject to a sonic object (a thing to be listened to) but rather can begin to develop an understanding of sound in terms of a corporeity which is always in the process of arriving at its own eventuality.

The body is never able to be thought in terms of a completion, that is to say, but always only in so far as it decays and develops – entropically, intellectually, psychically, physiologically, etc. There is never a moment at which the body is fully realized or finally formed, a point at which the processes of its own internal densities come to rest. No such moment short of death, when still the body carries on towards the point of a passing through into some other dimension beyond reach of a conscious 'self'.

And so for living bodies the recording studio marks a space in which it might be said that the body can experience its own death – externalizing itself as a prosthetic trace that memorializes a living presence no less assuredly than a gravestone. A recorded voice is always, then, a voice from the grave, from a beyond the body, even though it reaches the body as an event to be actualized (where there is both loss and acquisition) through a kind of endless corporeal receptivity. This is of course especially true of one's own recorded voice, which arrives in a form of the self as other, but is perhaps no less true of the recorded voice in general in so far as this must always take the form of a lost and open detachment, like a page torn out of someone else's diary.

No less than a dream, in other words, a sound is a corporeal event. This doesn't mean that sounds eventuate in terms of a sort of private interiority, either in a psychoanalytic or an existential sense. For this might be to suppose that dreams are the expression of an unacculturated unconscious or an individual's essential alienation. Dreams, although they happen in the body, do not occur outside culture.

This is why it seems to me that we can think the sound of a dream, on the model of being able to dream a sound. And there is no way of knowing whether the sound of a dream might be different from so-called other sounds that occur in our bodies while awake. But if sound can indeed be thought this way, as corporeal and dream-like, then it may be that as a medium (and I'm thinking here especially of sound-usage with regards to radio) sound can be understood in terms of accidental effects, to which it's no less prone than a body dreaming or awake.

Every body is an accident, waiting for the ultimate accident to occur – death. This too is part of our corporeality, so that perhaps in every dream and sound that eventuates in our living bodies we're reminded of that accident that remains to become, at the moment of our unbecoming, crossing to an afterlife like switching to another channel.

Yet for all that sound is nonobjectifiable, it should not be taken as an absolutely special instance of semiosis. As we'll see in the following two chapters, even a concept as seemingly assured as 'literature' is fraught with problems of objectification. And so if we cannot say with any 'scientific' certainty (despite earlier attempts on the part of semiotics to do so) what literature *is*, then we may well ask what kinds of possible statements *can* we make about 'it' and how might these affect the teaching of literature? For if a concept as seemingly fundamental as even sound can be opened to some troublesome thoughts, what sort of confidence can we afford to invest in the stability of such a culturally binding and reassuring concept as literature?

7. Catholic English

As a schoolboy at Christian Brothers College in Fremantle (Western Australia) in the 1970s, I studied English and English Literature. Although these were different subjects, it's easier now for me to say what English Literature was: lots of Shakespeare (plays and sonnets), plenty of poetry (some of it modern) and classic novels and short stories by authors whose names were seldom unfamiliar to me. Whether it matters that I read The Canterbury Tales *for my matriculation year because I attended a Catholic boys' school, I cannot now say; perhaps there were reasons other than the Brothers' defiance in the face of declining standards of education for the books I was given to read.*

I well remember the Brother who taught me English Literature. A tall, dandyish man of sixty with oily white hair and a ruddy complexion, he came to Fremantle in time for my last year at that school. Brother M. spoke several European languages, had a good deal of Latin and was the first real scholar I had ever met. He was also, at that time, the most arrogant person I'd ever met, and may still qualify as such. Of course, I was flattered that someone who so obviously did not suffer fools gladly seemed to suffer me. He encouraged me to read my work aloud in class, was always willing to stop and talk to me in the school grounds about the books I was reading, and wrote kind comments on my essays. Once, he even quoted a fragment of a Paul Simon lyric at an evening seminar between our Eng. Lit. class (there were seven of us, I think) and the girls from a posh convent school on the other side of the river, where the seminar was held, as an instance of great metaphor. He drove three of us there, and we were allowed to smoke in the car ...

When I think back on all the crap I learned in high school
It's a wonder I can think at all ...[1]

I suppose it must have come as a shock to Brother M., although it didn't to me, that I failed my matriculation. I don't remember what I got for English Literature, but it wasn't the Distinction everyone expected, while all I remember of the grades for my other subjects is that they weren't enough to get me into university. I probably could have joined the public service or gone into advertising or become a writer, but all I knew at the time was that I was a failure. I knew this because my parents were so ashamed. Personally, I still thought I was pretty smart.

My final year at CBC I spent growing my hair and partying. By the end of Lent I had the longest hair in the school, which is why I was called upon that Easter by our Headmaster to play Jesus in a lunch-time performance of the Stations of the Cross. We did it wearing Levis, everyone slouching through or mumbling out his part against a soundtrack of our favourite records. On my way to Calvary, dragging a huge telephone pole, I remember one of the 4th-Year students calling out that Jesus didn't wear a wristwatch! Then, a few months later, late one Friday afternoon the Headmaster told me to get a haircut by the following Monday morning, or don't bother coming back.

I got my hair cut (at my father's insistence), grew totally introverted, and failed my exams. The next year I repeated my matric at the local high school and went to university in 1975, where I failed my first year. I spent my English 100 exam playing pool and drinking at 'Steve's', as the local pub, the Nedland's Park Hotel, was called, because I'd never attended a tutorial in the course and didn't know what was happening. I'd been embarrassed by my surname and wasn't able to bring myself, at the start of the year, to sign on for a class – and I was too scared to withdraw. All I passed that year was philosophy; so I had to write a letter to the Dean explaining why I should be allowed to continue my university studies . . .

More or less, this has been the story of my life. I don't remember a time when I haven't liked books or, at any rate, print (I was a voracious reader of horror comics as a child), and I suppose that this eventually translated, under the Brothers' supervision, into my being able to write a passable sentence. Or so I like to think. To this extent I guess the Brothers did succeed – succeeded in nurturing in me the very mark of the educated citizen: the ability to write. *Whatever else I might do with my life, for however long I might continue to act the nonconformist, at least I know what I'm doing when I choose to split an infinitive. I know, at least, where to put the verb in an English sentence; and if that's not a matter devoutly to be wished, nor is it anything to sneeze at.*

Of course, it took some effort to end the previous sentence with a preposition (not quite of the order of having impure thoughts, when I was a child, but almost). So I confess it here, along with my other grammatical sins and those I am about to commit. It isn't easy to be sly when even what goes on inside your head is properly a public document in the making; all it takes to turn it into text is the memory of some priestly interlocutor behind a dark, drawn curtain of the mind.

Or do I, in fact, protest too much? For surely it is to overstate my offence (nowadays among the most venial errors of the language) if, by having ended a sentence with a preposition, I must immediately confess to a guilt which is then likened to a Catholic schoolboy's fear of his own sexual desires and curiosity. I have strayed, surely, from a consideration of things prepositional to things preternatural, from human to divine law. So my offence is surely a trope, a bathetic straining to confuse an internalized English *voice with an internalized* Catholic *one. While they both monitor my many and varied trespasses, surely they cannot both produce in me the same feelings of guilt.*

But the fact is I can't say, and this may be all I want to say here. I want to say that, for me, the history of studying English Literature is inseparable from the history of growing up Catholic. For me, the genealogy of English is Catholic through and through; it's why, even now, as a 'lapsed' English Literature graduate whose attention has strayed to questions concerning technology, the concept of culture and so on, I find myself wanting to write about teaching English . . .

I have taught literature courses at several universities, and this is my one desire: that students should learn to accept that James's governess is mad and, equally and undecidably, that she sees ghosts.[2] In my experience, this is for most students the equivalent of learning to accept the dimensions of a square triangle – it goes against everything they know, which is everything they've been taught. That is the crux; because not even their teachers have learned to accept this, as a serious possibility. With few exceptions, professional readers of *The Screw* have argued long and hard over the question of the governess's mental health, and thus, they suppose, the real point of the story.[3] To decide either way, however, regardless of how relentless or even convincing an argument is put for doing so, is at the same time to perform an act of faith, no less illogical than believing in the virgin birth.

By the same token, the reading of James's story that I'm advocating here is no less illogical itself. To read a story and not to know what it means, or worse, to read it in *order* not to know what it means or to read it *knowing* that what it *means* cannot be *known*, might seem to defeat the purpose of reading. Or it might mean that there is room for debate over, or manoeuvre within, what meaning means. To take an example: why shouldn't we believe that Mon Cul, the baboon, a character in Tom Robbins's *Another Roadside Attraction*, knows an English word that rhymes with 'orange', which is what he claims to know?[4] Is the fact that Mon Cul is a baboon, or that he refuses to disclose the rhyme to anyone, grounds for not believing him? After all, having accepted that Mon Cul is a *character*, at what point (and under what circumstances) could we then dispute his (or its) powers of knowing, depth of experience or whatever on the basis of comparison with what 'real' people (that is, not characters in fictional narratives) are known not to be able to know – such as what a square triangle looks like, or an English rhyme for 'orange'? Let alone that in real life baboons, as opposed to people, don't, we suppose, know anything remotely of the sort. (My arse, they don't!)

If we're willing to accept that the mathematical symbol π has no real referent outside a highly specialized semiotic system, why should we seek automatically to impose an 'outside'-logic on a fictional character?[5] Indeed, in real life we believe all manner of nonsense to be true: we believe, or some of us do, in the historical, theological or preternatural truth of at least one virgin mother, for instance; while others believe that Elvis isn't dead, ghosts are real, God invented the AIDS virus and the Martians have already landed, etc. Nor do these beliefs result necessarily in their believers becoming socially dysfunctional, as proven by the many members of the business, political, arts, intellectual and other communities who, despite their spiritual, fantastical or otherwise 'illogical' convictions, nevertheless continue to hold down important jobs or to do important work.

But it is decidedly more difficult to hold down an important job, within a university English department, if what are judged to be one's beliefs are seen also to be contrary to those of the discipline's canonical truths. Hence *either* the governess is mad *or* she sees ghosts. Isn't this what English Literature should be able to decide, on the model of some principle of reason underpinning the formation of subjects taught at university level? For can it be allowed that English teaches something else, something not quite of the order of reason at all and which has to do with the formation of a human subject rather than an institutional one? Accordingly, and here's the rub, this subject could be appreciative and still be professional.

If it were the case, in other words, that what English Literature teaches is the self-enabling capacity to appreciate a certain set of writing practices as the raw material of literature and to evaluate individual performances of and within those practices, then surely it must follow that the professional *littérateur* is an appreciative human being. Rather than reading simply for pleasure, as the common reader might be said to do, the professional reader reads for the more transcendental purpose of *appreciating* what is being read.

To be sure, while appreciation may take several forms, including, as the dictionary puts it, 'the act of estimating the qualities of things according to their true worth', it is surely a strange professional purpose on its own. Under the surgeon's scalpel, for instance, I would prefer to be operated on by someone whose practical skills, rather than his or her sense of appreciation, got them the job, although I would certainly expect their judgement of the condition of my internal organs 'according to their true worth' to number among such skills and to count as professional knowledge.

The interesting thing with regard to this analogy, however, is not so much what counts as professional knowledge, but what counts as 'true worth'. This is interesting because 'true worth' is not the same in both instances; the true worth of my kidneys, for example, is relative to my overall physical health, and has very little to do with kidneys in their ideal form. So the true worth of my kidneys is a matter of practical judgement (or appreciation), albeit the practicality of the matter having to be decided by a professional. Always, that decision is a contingent one. But this is not what the 'true worth' of a work of literature means among professionals who appreciate it, and who appreciate it precisely for the *non*-contingency of its 'true worth', essential nature or ineffable literariness. The true worth of a work of literature, then, seems not to be a matter that remains to be decided or judged – its status being always already given.

Now it might be argued that this is true also of human kidneys. In other words, at any given time there may be debate among professionals as to the true worth of my kidneys in relation to my general well-being. If, for example, I were a known drug-abuser, then the weight of medical opinion might favour the view that my kidneys were in remarkably good shape, under the circumstances; or it might subscribe to the opinion that they were typical of my condition; and so on. But whatever the view, the 'true worth' of my kidneys would be decided by professionals who were empowered to make decisions, to make meanings. Needless to say, this is not how the professionals themselves would understand their role; for them, it would be a matter of examining my kidneys for signs of their 'true worth', and not a matter of making it up. Always, for the professionals, the true worth of my kidneys would already be decided in advance – by nature, as it were – so that their professionalism would be called upon to make not a decision, but rather a discovery.

This, then, is the essential nature of 'true worth', the true worth of 'true worth' itself. Once it has been decided, in other words, 'true worth' is immediately reconstituted as a discovery, an always already there, prior to suddenly being noticed for the first time. Like God, DNA or 'Newton's' law of gravity, it exists without us. Just as now we're beginning to discover that chaos has existed without us, too.[6]

On this model, therefore, it must be in the nature of truth that something must already be true *before* it's understood to be so, or before its truth is able to be appreciated. The true worth of my internal organs *and* the true worth of a poem cannot both be 'true' *and* remain to be decided. In this sense, admittedly a seemingly idiosyncratic one, 'true worth' is always and strictly what is *undecidable*. Like God, if you believe in Him, or equally even if you don't. By the same token, 'true worth' is not always absolute (although it's always assumed to be a priori); more often than not, indeed, its nature is contingent. The difference between an appreciation of my kidneys and Milton's 'Lycidas', say, depends on knowing that my kidneys, like the rest of me, must perish; because they must go the way of all flesh, their 'true worth' is never absolute but remains in flux, subject to contingent forces in the form of my age, general health, lifestyle, physical environment, etc. Any appreciation of the true worth of my kidneys must, then, be an act of discovery which is temporal, contingent and ulterior – and is therefore of the order of a utility. On the other hand, the true worth of 'Lycidas' must be absolute and everlasting, though Edward King himself be dead.[7]

If the category of *absolute* true worth were not available to literary scholarship, of course, there could be no canon. Nor could we have invented God without first (or afterwards) positing the prior existence of His absolute priority to our decision to invent Him, thus transforming our decision into a discovery of the truth. Otherwise, we have to believe simply that God invented us. Or, as Nietzsche writes, concerning the imperative priority of God's existence to His creation by the Jews:

> *Race* is required for it. In Christianity, as the art of holy lying, the whole of Judaism, a schooling and technique pursued with the utmost seriousness for hundreds of years, attains its ultimate perfection. The Christian, that *ultima ratio* of the lie, is the Jew once more – even *thrice* more ... The will to employ as a matter of principle only concepts, symbols, attitudes manifested in the practice of the priest, the instinctive rejection of every *other* practice, every *other* kind of perspective in the realm of values and practical application – that is not tradition, it is *inheritance*: only as inheritance does it have the effect of a natural quality.[8]

But if not race or ethnicity, then certainly a particular force of will. A force that, I propose, having invented God or English Literature, there-after presides over its creation with the vigilance of a priest.

Or, to put this rhetorically: what cause for vigilance otherwise, if the 'true worth' of the canon were not indeed 'a natural quality' of the canon 'itself'? If the Great Tradition were, in fact, a *tradition*, formed out of some contingent historical perspective, say, or fashioned from within a collapsed ethico-technical domain of praxis ('the realm of values and practical application'), and if it were understood in these (politico-historico-philosophical) terms, then it simply could not be understood also, in theological terms, as canonical.[9] Quite undecidably, in other words, it is both exemplary of a tradition and also exemplary of a naturally occurring (or divinely sanctioned) phenomenon. In this sense it is passed on not simply like, but literally *as* an 'inheritance' (a redemptive one according to Prof. Gordon, as we may recall from Chapter 1). The canon's dissemination, that is to say, depends on the insemination *by* the canon of each newly great scion that can trace its lineage from *Beowulf* to (and, still, not much beyond) Virginia Woolf. Such a form of in-breeding, of course, produces a version of genetic purity which threatens the breeding stock with its own extinction, though not before it has been threatened with going insane. Hence (or perhaps) the priestly rancour from among the guardians of the canon against what they might be seen to see as Modish Literary Apocrypha (MLA).

But let us turn to an example. The following passage is taken from an essay in a special issue (on literary theory) of the European journal *Poetics*. Its authors, Richard Freadman and S. R. Miller, following a discussion of 'F. R. Leavis and Terry Eagleton as exemplars of drama-tically opposed views of "literary theory" (also as exemplars of the two "paradigms" discussed)', conclude:

It hardly needs saying that the account offered here of literary theory and theorising is both preliminary and extremely limited in scope. However, the intention of the discussion has been to sketch in a 'third' position on literary theory which advocates the following: one, that theorising has a necessary and legitimate part to play in literary studies; two, that it needs to be guided by objective principles; and three, that texts and their interpretations will always to some extent exceed the explanatory capabilities of theories. That texts do so exceed the grasp of theories is in part a consequence of the fact that interpretations rely to some degree on the experiential powers of readers. We conclude, therefore, that a residual subjective element will inevitably be present in all interpretations.[10]

Would it, however, be an observation 'guided by objective principles' to note that Freadman and Miller dispense with the inverted commas that surround their initial use of the term 'literary theory' when they use that term again for the last time on the final page of their article? Would it be professional or perverse to ask – is this the same term on each occasion? Indeed is their 'literary theory' or their ' "literary theory" ' the same as our MLA above? Or would this line of questioning be an instance of the sort of theorizing that does not play a 'necessary and legitimate part . . . in literary studies'?

What kind of theorizing does play such a part, then? In a word, what do 'necessary' and 'legitimate' mean in relation to the professional rules and procedures of literary criticism? For surely the kind of professional knowledge invoked by Freadman and Miller is underpinned by a notion of the consensual limits circumscribing the acceptable or appropriate conditions of knowing or discoursing on, in this case, the literary text. According to this knowledge, such a text is whole, unitary and self-identical. More or less, that text is the book. So long as agreement remained on this most important of literary critical facts, then even forms of criticism which were (and continue to be) committed to connecting a single work of literature to a single life's work, a literary genre, period or movement, or some other structure of continuity, but which still held (or still hold) to the *integrity* of the single text, could continue to flourish. On this one fact, at least, even the Leavisite and the semiotician could be in accord. Or as Foucault (quoting Althusser) writes, following a synopsis of the shift within disciplines of knowledge from structures of periodization to an understanding of rupture:

> The most radical discontinuities are the breaks effected by a work of theoretical transformation 'which establishes a science by detaching it from the ideology of its past and by revealing this past as ideological.' To this should be added, of course, literary analysis, which now [in 1969] takes as its unity, not the spirit or sensibility of a period, not 'groups,' 'schools,' 'generations,' or 'movements,' nor even the personality of the author, in the interplay of his life and his 'creation,' but the particular structure of a given *œuvre*, book, or text.[11]

In other words, despite what traditional or conservative literary criticism regards as semiotics' heretical desire for a science of the text, a science of literature, structuralism and semiotics remain committed to a philosophy of grounds, of integrity, and even to a species of origin. For *something*, for semiotics, grounds the text and lends it self-identity, and may be traced therefore to a beginning: if not the author or his or her community, or moment in history, then *structure* is what is held to hold the text together, albeit in ways which are not always obvious to the professional let alone the common reader.

Any debate between structuralist and traditional forms of literary analysis, then, must be internecine to some degree. Quite simply: the most deeply structured text is still, if one's ideology requires such an attribution, attributable to the genius of its author. Once it is possible to retain authors and masterpieces (even in the form of 'deeply structured texts'), of course, so it is possible to save the canon, and thus one's profession or calling. (Similarly, it was once considered prudent to bend with the wind and change the Catholic mass from Latin to living languages.) This can be done in the name of a certain catholicity of reading that enables the sanctity of the literary text to remain the miracle of composition it has always been.

Appreciationism, as Freadman and Miller clearly seek to persuade themselves, while it may be threatened by analytical approaches to the literary text, is nonetheless capable of being argued back into a position of critical pre-eminence, as the governing condition of what constitutes 'necessary' and 'legitimate' critical practice in the service of interpreting the true worth of the text itself. Hence no loss of faith.

But what if there is no 'itself' of the text? Then there would be crisis – of a kind exceeding the hope of reinscription into understandings of the text as sacramental host.[12] Then there could be no parasitical readings, nor sacrilegious or heretical ones. There could be no such readings, for there would be nothing 'there' to 'read' or 'appreciate' *in the first place*, prior to being read. But this is not the same as grounding reading in the 'experiential powers' of readers, whose experiences of life and literature must be different. In practice, at any rate, Freadman and Miller's gesture to the indeterminacy of the literary text, based on its interpretative infinity (or indefiniteness) according to an indefinite variety of readers' stockpiles of experience in which its meanings are grounded, is simply not what passes for literary analysis in the public domain. Certainly degrees of difference are allowed among interpretations of a work of literature, but seldom differences of kind. So, for example, if I were to maintain that *Wuthering Heights* is no more 'literary' than a Mills and Boon romance, my reading would be outlawed – on the grounds that I was untutored in the skills of literary appreciation, perhaps, or that

I was being mischievous but, in any case, plain wrong. If, on the other hand, I were to offer a reading of Emily Brontë's novel which compared the incidence of typographical 'errors' between, say, the Norton and Penguin editions, I would doubtless be accused of not 'learning anything' from the text, or simply (ab)using it as an *exemplar* of my 'critical presuppositions'. Again I would be *wrong*, but this time as an *atheist* is wrong – out of choice, perversely. And so while a pagan might fail to appreciate the true worth of *Wuthering Heights*, at least he or she can recognize that it belongs to a certain set of writing practices constituting the raw material of literature. The pagan's problem is at the level of an evaluation of the text's performance of those practices which, out of ignorance or mischief, the pagan doesn't rate very highly. Only the atheist, however, can deny the very existence of the first condition of literature's true worth, and hence the true worth of literature itself, displacing the need for evaluation altogether. On the other hand, what is seen by believers to be an act of denial on the atheist's part is not how the atheist understands it. From within an experience of faith, in other words, atheism is always a *denial* of that experience. Of course, what this presupposes is the prior and eternal existence of a God for the atheist *not* to believe in.

'Atheism', then, is determined by a discourse that posits God as the living ground of human existence, with the promise of eternal life for those who acknowledge that ground and show devotion to it. 'Atheism' is the term for those who are not devout; it has meaning on the basis only of its opposite's condition as the proper or truthful state of affairs, for which there is no single term. If 'atheism' were the proper condition of human existence, however, as it is understood to be by 'atheists', then *its* opposite would be the special case, for which there would certainly be a term of designation.

By a similar logic, literary appreciationism is able to stand firm against the critical perversity of a species of reading which refuses the priority and integrity of the text. But which 'text', exactly? The one with, or without, typographical 'errors'? And how would we recognize such an instance of 'error'? And why would we want to exclude such instances from the domain of 'legitimate' critical observations? And who would decide these questions?

Clearly, in practice, the decision could not be left to the 'experiential powers' of readers, whose indefinite variety of experiences would produce an indefinite set of answers and thus no answers at all. Unless, however, we were to suppose that human experience has a common ground. In this case, even allowing for cultural and historical differences, the range of actual differences among readers' interpretations of literary texts, based on their experiences, which would be grounded in 'an' or 'the' experience, would be of the order of degrees. And this is the humanist loophole, of course, enabling the conservative critic to argue the 'limits' of pluralism. Thus M. H. Abrams can write that, '[a]s a critical pluralist, I would agree that there are [sic] a diversity of sound (though not equally adequate) interpretations of the play *King Lear*, yet I claim to know precisely what Lear meant when he said, "Pray you undo this button." '[13] Moreover, he can presume with confidence that others will also claim to have access to this knowledge, since 'language ... is a cultural institution that developed expressly in order to mean something and to convey what is meant to members of a community who have learned how to use and interpret language'.[14] So the short answer to the question of who or what determines what constitutes, in the first instance, a 'text' and then 'legitimate' critical understandings of 'it', is something like *the skilled members of an interpretative community*.

Just as everyone is born with a potential to know God, according to the faithful, so Abrams would have us believe that we're all born with a potential understanding of the line he quotes from *Lear*. In order to realize this potential, of course, we must undergo instruction by learned hermeneuts in the proper ways of interpretation, and the limits pertaining to these ways; having acquired this knowledge, we become members of a community – whether of redeemed souls or educated readers, or both. Surprisingly, or (un)fortunately, though, Abrams' example is ill-chosen. For without even needing to have ever read or seen a performance of *King Lear*, the line whose meaning Abrams claims to know beyond all equivocation is easily shown to be equivocal of itself. Indeed, its undecidability is a matter of public record. Here I quote from Kenneth Muir's gloss on the line in the Arden edition of the play, where the following entry appears:

> Lear feels a sense of suffocation, and imagines it is caused by the tightness of his clothes. J. W. Harvey suggests that Lear is referring to one of Cordelia's buttons; but I think this is unlikely. See letter in *T.L.S.*, 14 Nov. 1952, and later replies.[15]

The full text of Lear's speech containing the line is as follows:

> *Lear*. And my poor fool is hang'd! No, no, no life!
>   Why should a dog, a horse, a rat, have life,
>   And thou no breath at all? Thou'lt come no more,

Never, never, never, never, never!
Pray you, undo this button: thank you, Sir.
Do you see this? Look on her, look, her lips,
Look there, look there!   [*Dies.*]

(V.iii.304–10)

From these passages it becomes clear that there are several inconsistencies between what Abrams claims to know and what others have
claimed to know as the line's meaning. Most remarkable is that Abrams'
version of the line as it appears in *Critical Inquiry*, 'Pray you undo this
button', does not include a comma after 'you', whereas in Muir's edition
of the play a comma does appear at this point. Without entering into a
dilation on typographical errors, we must notice (even if only to be
pedantic) that this inconsistency throws into doubt the very status of
'the' line itself, a status that surely must depend on the absolute
knowledge of which line is not Abrams' or Muir's, but William
Shakespeare's. In other words, if it cannot be decided which line is the
original, then it cannot be decided whether Abrams or Muir is 'misquoting' – so that effectively they both are. And if we can't decide this, how
can we decide a *meaning* for 'the' line?

But who would have thought a comma to have had so much
indeterminacy in it? Or, in other words, surely we must be guilty of
murdering the line at the expense of 'learning anything' from it, despite
having shown that there is no certain 'it' of 'the' line, or the line 'itself',
except maybe from within the realm of common sense, where it can be
readily agreed that the difference between Abrams' and Muir's versions
is inconsequential to determining *its* meaning. Yet not even this is
certain, as the correspondence in *The Times Literary Supplement* serves
to document. On the other hand, perhaps the matter can be decided by
referring to another edition of the play, where, if we were to find an
obliging comma at the right place, we might reasonably conclude that
Abrams is guilty of misquoting or *Critical Inquiry* of committing a
typographical error. Thus I cite *The Illustrated Stratford* edition (the only
other Shakespeare to hand as I write) of Lear's speech above:

Lear. And my poor fool is hang'd! No, no, no
          life!
Why should a dog, a horse, a rat, have life,
And thou no breath at all! Thou'lt come no
          more,
Never, never, never, never, never!—
Pray you, undo this button: thank you, sir.—

Do you see this? Look on her,—look,—her lips,—
Look there, look there!—   [*Dies.*]

                                                         (V.iii.306–12)[16]

Clearly, the comma is in place. But still there is a question: which
place? For in the Arden edition the comma appears at line 308; in the
Stratford, at line 310. Moreover, the Arden's capital 'S' for 'Sir' (l. 308) is
replaced by a lower case 's' in the Stratford (l. 310), while the Stratford's
several em-dashes (ll. 309–12) do not appear at all in the Arden edition;
and of course the two versions of the speech are set differently on the
page. Normally, of course, such information would not be accredited
much interest in the field of Shakespeare studies except with reference
to editorial choices that have to be made on the basis of grammatical and
other inconsistencies between folio versions, etc. While these are seldom
determined in an absolute sense, they are made *sufficiently* determinate
for the proper business of Shakespeare studies to stay under way, that
business being the exegesis of the plays' meanings rather than the
grammar or the science of their individual composition. And so
although we can show that there is no *absolute* original Shakespeare
text (each play, for instance, having at once several 'slightly' different
editions to 'quote' from), there is always a *sufficient* 'original' upon
which to do 'legitimate' and 'necessary' exegetical work. For as long,
then, as the business of Shakespeare studies remains a hermeneutics of
the text, interpretations can be worked out and agreed upon which are
for all intents and purposes absolute although in fact sufficient. Thus it
is a virtual absolute fact that Lear is referring to his own clothing and not
his daughter's when he says, 'Pray you undo this button,' or 'Pray you,
undo this button'. That is to say, it's a sufficient fact in so far as a
majority of members of the community of Shakespeare scholars have
agreed that this is what the line means in its most basic referential sense.
It's a sufficient fact because it enables the most widely accepted, or most
widely practised, reading of the play to continue being (re)produced,
which is that Lear comes to know himself only after being disrobed of
the trappings of office, in the first place, and thereafter of human society
or civilization. So it fits this reading of the significance of Lear's
nakedness to his own self-understanding, as much as to ours, that one
of his last gestures (and perhaps his final act of metaphor) should be to
loosen his clothing before he dies. Having made the line consistent with
an exegesis of the play, it can then be attributed to an absolute source or
origin: the playwright's intention. This then is what Shakespeare must
have meant, because this is the most *likely* meaning of the line (based on

its sufficiency, of course). Therefore there *is* an original of the play, although it is not absolutely absolute, which has its source in the intentions of the one who wrote it, those intentions being 'there' *in* the miracle of the text itself regardless of where an editor might choose to put the commas, or even if there is debate over the inclusion of several lines, etc. More or less, then, punctuation marks cannot be considered, unless by editors (and then only under certain conditions), to form an *organic* part of 'the' text, whose integrity cannot be compromised by grammatical niceties or typographical errors.

So, for example, Melville's short stories are still 'Melville's', even though their author is on record as having left their punctuation to the discretion of his publishers of the time (for *The Piazza Tales*) and thereafter to successive editors of his work.[17] So too are F. Scott Fitzgerald's novels and short stories his 'own' even though he was notoriously lax in his spelling and punctuation, leaving his editor, Maxwell Perkins, to correct these in the manuscript before publication.[18] Routinely, in other words, the work that readers (common or professional) read is not precisely what's been 'written' (or submitted for publication) by the person whose name appears on the cover of the published book and who is celebrated as its author. The books we read, although never absolutely 'the' original work of an author (if only in so far as we read them in a published font other than the one in which they were 'written'), nonetheless are sufficiently indistinguishable from 'the' original work to count, effectively, as such. More importantly, they're sufficient to enable literary criticism and the English department, in their conservative forms, to go on living on. It is only when the sufficiency of these texts, upon which these forms are dependent, is called into question that the conservative department and literary studies are in danger of collapse.

In the face of such a threat, there is only one recourse: namely, to 'common' sense. Only common sense can set the limits of what is 'necessary' and 'legitimate' in terms of reading practices, if those who actually determine the necessity and legitimacy of specific reading practices are called to account for their defence of these historical contingencies. Hence Abrams' recourse to a theory of language as 'a cultural institution that developed expressly in order to mean something and to convey what is meant to members of a community who have learned how to use and interpret language'. As we have seen, however, this appeal to the 'obvious' results in indecision as to the very necessity and legitimacy of a single comma; how, then, can it hope to decide the meanings of larger texts? Quite simply (or commonsensically), it must do so by denying that any indecision over a comma is properly a matter of concern to the purpose of reading, which is to understand, to get at, the meaning of the text, to get at or to appreciate its 'true worth'. For it is *the text* which is held, commonsensically, to be self-evident, whereupon the purpose of reading is obviously to understand it. Yet there could be nothing *to* understand (or something else to understand, or another form of understanding is required) *if there is no text.*

How could there be, if different editions of what is called 'the' text can appear with different punctuation marks and still be 'itself'? So what is called the text, in other words, is itself caught up in the play of textuality, the deferral of its own self-constitution. Nor is this the case only with 'special' instances of the text, such as *King Lear* or one of Melville's stories, where there's room for play in deciding where to put the commas and how many there should be. For these 'special' or 'unusual' (or 'historical') instances are in fact wholly typical of the text (itself), in so far as they throw into question the premises by and on which the norm(al) text is understood. It must be able to be decided, in other words, what *is* 'proper' to the text's self-identity if it can be decided, commonsensically, what isn't (punctuation marks, typographical errors, page layout, typeface and so on). If there is a domain of the im-proper (the unnecessary, the illegitimate), surely there must be a decidable domain of the proper; otherwise, of course, *there is no text.*

But what would proper data of the text itself look like? If it must be a domain which doesn't comprise signs in their materiality, might it do so according to their appearances? And who would decide the correct or proper interpretations of these, if not the priestly critics? One has to maintain simply, in other words, that the 'true worth' of the text is uncontaminable by effects of improper analytical methods (for example, MLA) or by improper typesetting. Similarly, one may defile the eucharist but never destroy its sacramental nature, its 'true worth', for this is uncontaminable by human perversity (or perversion). Each host, as it were, is utterly incorruptible, pure beyond adulteration, and identical with itself in all its forms of material variance. Understood in these terms, the question of whether Abrams or Muir has 'misquoted' Shakespeare is decidedly improper, for the 'true worth' of the text transcends this quibble. Again this does not mean that 'the text' is absolutely absolute, but simply sufficient for the purpose of understanding 'it'. Had Abrams 'quoted' the line to read, 'Pray you undo this gown', for example, there's little doubt that we could accuse him of 'misquoting', without any need of inverted commas around the charge.

In the form in which the line appears in *Critical Inquiry*, however, there is sufficient doubt to allow that the journal may be guilty of a typographical error rather than Abrams of a bibliographical one, so nearly is the line a perfect quotation (at least on the evidence of the Arden and Stratford editions of the play, as long as we agree to overlook the problem of the line's proper place). But even if it were to be decided that the line as it appears in *Critical Inquiry* is, strictly, a misquotation (and so it would be, regardless of being a typo, if either the Arden or Stratford edition were taken to be the source), it need not be taken to have any sufficient effect on Abrams' claim to know the 'true worth' of its meaning. This would be less certain, though, if the line had appeared in the form of our deliberate misquotation above, for then the line would be sufficiently different from the source as not to be self-identical with it; effectively, it would be absolutely different.

The question must be asked: why this difference between venial and mortal forms of misquotation? Why is a misquotation that forgets a comma effectively not a misquotation, while one that replaces a word absolutely is? Presumably the answer lies in degrees of violation of the 'true worth' of the text; although this is always ultimately inviolable, certain acts of defilement can nonetheless be wrought upon it. One such act, clearly, is the replacement of an 'original' word, which is passed on (or passed off) as an accurate quotation. This might be said to be an act of hermeneutic violation. On the other hand, it is not an act of violation to forget a comma, since this violates (and then only venially) what might be called the *incidental* integrity of the text, in the sense that punctuation marks are scarcely more organic in terms of the text's 'true worth' than the font in which the words appear. What this means is that, despite his act of forgetfulness (if that is what we choose to call it), Abrams has, in fact, *quoted* Shakespeare in the line that appears in *Critical Inquiry*. The line *is* a quotation because it preserves the hermeneutic integrity of Shakespeare's words (although not, as I hope to show, his 'writing'), and is therefore self-identical with the 'true worth' of the original. What has been quoted, then, is not the line's materiality (which can never be quoted, absolutely) but its appearance. What's been 'quoted' is the line's *true worth*.

But surely this must be a trick. Because if we can forget the comma and still be allowed to have quoted the line, then surely the 'true worth' of the line is not actually 'in' the line itself; in which case, where is it? And if it is not 'in' the line, then isn't it also not 'in' the writing? In which case, where is it? And surely it can't be 'in' the writing, because we can be said to have quoted the line without having properly punctuated it (according to the accepted, written or published, form of its source). So where is it?

If not in the writing, then perhaps (to 'quote' Derrida) in (the) 'speech'? By this I mean in the domain of speech's metaphysical priority to and over writing (see Chapters 3, 4 and 5). Let us take an example. Here I type a word in 12-point Palatino (a Macintosh font, presently installed in the system memory of my computer):

button

Here is that word in 14-point Geneva:

button

Here it is in 18-point Courier:

button

Now it is clear that in their materialities these are different signs. But I have been calling them the same word, 'button'. I can do this only by overlooking their differences, which are effects of writing and *cannot* be spoken. How does one speak 'button' except as 'button'? Nonetheless, I repeat, these are different signs, although in the strictest sense they are the same word. (That is, the word 'button' remains the *word* 'button' regardless of how it is written.) Since I have – *in writing* – referred to these different signs as the same word, and since it is unlikely to be objected that there's anything unusual or improper in my having done so, we may thus assert the following: 1) signs become words on the basis of their speech-identity; 2) the speech-identity of a word has precedence over its available forms of writing-identities. This is to say that the written form, 'button' (or 'button'), is understood to be self-identical with the spoken form of that sign (namely, the word 'button'), which is a quality of the sign that remains inviolable to effects of (mis)representation, or immune to writing. In short, writing is a technology for copying speech. Or, in terms of the present argument, it's a technology for copying the sign's 'true worth'.

Strictly, we cannot speak the sign ',' although we can speak the sign 'button'. So the former must belong to the domain of pure writing, whereas the latter derives from that of 'pure' speech and is thereafter (according to common sense) copied into the domain of writing. Thus the difference between 'button' and 'button' *is* purely a writing difference, since it has no speech equivalent, but is nonetheless and therefore a difference of a purely technicist order. As such, it is not a difference which has been of any interest to literary studies in its conservative forms, where the interest lies in the self-identity of words and not in the alterity of signs.

Yet this is remarkable. For, despite such insistence as that of Freadman and Miller on the inexhaustibility of the literary text, that insistence is grounded in a theory of speech's priority to writing which denies the word's indeterminacy and therefore allows for the possibility of sense to be exhausted by interpretation. This would seem to be an in-principle of the metaphysical priority of speech to writing; otherwise, how could M. H. Abrams claim to know the precise meaning of a line from *King Lear*? Hence it would seem, paradoxically, that there are *limits* to what is meant by the text's 'inexhaustibility' or its unlimited capacity to 'exceed the explanatory capabilities of theories' – unless of course the 'theory' in question were understood as the 'disinterested' and 'commonsensical' one of speech's privilege over writing. Only this 'theory' can, while never fully explaining the text, still appreciate its 'true worth' and the limits of its excess.

God works in mysterious ways, in other words, but this goes to show only that God exists. So it is with the pluralist, the catholic textualist whose tolerance for difference does not extend to *différance* – for the existence of the text itself is simply the first condition of the many readings that can be got from it. But 'many' must be kept from becoming 'any', or there will no longer be grounds for doing criticism. These grounds will always be the priority of word to sign, as a form of the speech/writing opposition, so that the text in its 'written' form will always be a *sufficient* approximation of its absolute 'true worth' in the form of 'speech' even if the written form contains typographical errors and regardless of how it is printed. Indeed, there can be a printing *error* on the basis only of a slippage in the writing up of the text's 'true worth', in its being copied out of the domain of speech and into that of writing. Only on the basis of a *proper* form of the text, in the realm of appearances, could we identify an errant form in the domain of writing and the sign's materiality.

If the text is effectively a product, therefore, of a theory of its ideal form as speech, then 'it' is in fact held in place by the textual practices and assumptions of that theory. 'It' is in fact a product of (more) text, since there is no place outside the theory in which 'the' text can be located. In fact, 'it' is locatable only as and through text(s). Any 'discovery' of the text's 'true worth' is therefore necessarily an outcome of *decisions* concerning the possible full range of things that might be said about the text, a range which is in fact beyond limitation. So when

Abrams claims 'to know precisely' what is meant by a particular text, he is in fact claiming to have *decided* what that text means on the basis of knowing what the text in general means; he has, in fact, turned a particular textuality *into* textuality in general (or simply into text), having interrupted or arrested a particular text's capacity to go on meaning beyond his claim to have arrived at a precise or final interpretation. Of course, this claim was absolutely true at the time that Abrams made it, in the sense that any *claim* 'to know precisely' the meaning of any text is absolutely true. But it's not true as knowledge simply for *being claimed* to be true, although it may well be sufficiently 'true' in the sense of being taken up or functioning as 'truth'.

But this is not what Abrams claims. He is not claiming a *situated* knowledge of a text; he is claiming rather 'to know precisely' what that text means. He is claiming *absolute* knowledge of the text, despite our having shown that there are several forms of that text and therefore no possibility of 'the' text at all. Again, as a claim, this is absolutely and irrefutably true. It is *true* that M. H. Abrams claims *to know precisely* what the line, 'Pray you undo this button', means. At any rate, it's true that he *claims* to claim to know this, or that he did claim to in *Critical Inquiry* in 1977. But so what? What can this tell us about the text he claims to know that isn't in fact more text, if only in the form of an interpretation that, as an interpretation, remains open to debate? For there to be *more* truth to Abrams' claim than the claim itself, there must be *a* text to be known. In this case, as we have shown, such a text (which would have to be absolutely self-identical in all its versions) is nowhere to be found except as an ideal of its 'true worth'. Neither does its 'sufficiency' make it into such a text, its sufficiency enabling merely the 'truth' of Abrams' claim to claim to know the 'truth' of the text. Beyond this, the truth of Abrams' claim is itself (more) text – and so on.

Now: how do you teach this?

In the Preface to his *Room for Maneuver: Reading (the) Oppositional (in) Narrative*, Ross Chambers writes:

> The issue of how to change the world is obviously at a rather remote horizon of the question I focus on: 'what happens when we read?' But I would not have undertaken this study if I had not thought the one was relevant to the other. My concern is not with what literature *is*, but with what it can *do*, and beyond that with the conditions of possibility that constrain what it can do. What it can do, I suggest, is to change desire; 'reading' is the name of the

practice that has the power of producing shifts in desire; and desire does not just produce 'fantasy' but reality itself. That is why a rather technical study of reading as an oppositional practice can hope to have something to say about changing the world.[19]

I'm all in favour of a 'technical study of reading' (a science of the text, as it were); at least this would make it a lot harder to take too seriously the mannered confessions of a teenage Christ impersonator, whose genre of writing is just another public staging of the self – a self trained to 'do' being an English graduate (who has already been trained to 'do' being a writer) and a lapsed Catholic. Piece of cake. 'Technical' studies of *reading*, that is, are a counter to literary appreciationism, which has nothing whatsoever to do with changing the world. Why should it? It's in control of it.

It knows, beyond a shadow of a doubt, that either the governess is mad *or* she sees ghosts. But this 'knowledge' is nothing more than an investment of belief in a 'myth' of consensus, which is still, as Peggy Kamuf puts it, 'dear to the most alarmist critics of literary theory'. In its certitude, the myth 'dissimulates or denies the conflicts that have never ceased to divide and redivide the ground upon which literary study is instituted'.[20]

How then do we work towards making another world, where *both* these statements about the governess are equally and oppositely true – and no one is bothered by this? No one wants to damn our souls to hell for even thinking so?

This is a question we will need to pursue in the following chapter.

8. Derrivations: From Derrida to Empson

It is common to suppose that with the spread of Derrida's influence in the 1970s the history of the discipline of Anglo-American literary studies underwent a break. But in order to suppose there was a break, the history of the discipline up to that time has to be imagined as continuous. In order to suppose that something dramatic happened, a certain historicity is required. Against such a background, the spread of Derrida's influence can be seen as an event, and one that threatens the very continuity of literary studies as a discipline by urging it to ask different questions and to develop a different analytics on the basis of having to rethink its fields and objects of knowledge. Since Derrida, or so a story goes, literary studies has been rent asunder by the struggle to define its disciplinarity in terms of whether it is proper to assume a specialist knowledge of texts-in-particular or, perhaps more speculatively, or more transcendentally, to define the discipline in terms of the very theoretical nature of the question of the text-in-general. However, and perhaps more radically, it may be possible to imagine another – discontinuous – history of the discipline, such that the event of Derrida's influence on literary studies may not be of the order (as an event) of something completely different.

Everything derives, though it is not always easy to say where anything comes from. No analytics of any state of affairs can hope to account for every determination or accident by which that state came about, let alone where it might be headed.

Loosely or directly, this statement may turn out to come from Derrida. In that case its locus could be the principle of 'adestination', developed in *The Post Card*, according to which every text must arrive at a destination but not necessarily at an intended one (see Chapter 5). So it would seem that 'Derrida' is able to function as a proper name to which the discovery of a principle can be attached, after the fashion by which scientific discoveries are credited – despite, in this case, the nature of that principle being that no such assurance is able to be guaranteed. No less than any arrival, every departure is an accident. Hence not even 'Derrida' can function as a fixed point within a system whose very functionality depends on the actual undecidability (despite the actual need for decisions) of every point within it.

This does not have to mean that ideas fall from the sky. There is no question that Heidegger, for example, is crucial to Derrida's discovery of the principle of adestination.[1] But if that principle carries any force, then it must be insufficient to attribute its discovery to a determined product of the history of a certain reading on the part of its discoverer. In this way (as it were) it may be appropriate to describe the adestinal formation of any idea by the term 'derrivation'.

However 'properly' Derridean it may be to pun on words, 'derrivation' is not confined to an instance of word-play for retaining a trace of Derrida's name. In excess of this function, although not contrary to it, derrivation refers to an asystematic system of (direct and indirect) 'influences' from within which texts are able to be positively identified as self-present. For example: what is still often referred to as a 'radical' approach to literature, deriving from the influence of Derrida's work on Anglo-American literary studies in the 1970s, may not be quite so new. Its derrivations may include what might be called nonpositive or immanent conditions of emergence. If so, of course, this remains possible only because Derrida *has* arrived as an influence on literary theory. Any derrivation, then, that might find immanent forms of Derrida's work in texts that were not directly influential on its development, would still belong to that work and would still count as a so-called radical approach. But what was understood to be 'radical' about the approach might have to undergo a shift in meaning.

What follows is a derrivation of an approach to literature that has come to be associated with a radically sceptical (if not a 'derrisive') attitude to questions of truth and actuality, which is attributed to Derrida's influence on literary studies. This implies that, before Derrida, no such approach was possible. In the following derrivation I wish simply to suggest that, albeit after Derrida, this may not be so.

In the Preface to the second edition (1947) of his remarkable and once widely read *Seven Types of Ambiguity*, first published in 1930, William Empson writes:

> If critics are not to put up some pretence of understanding the feelings of the author in hand they must condemn themselves to

contempt. And besides, the judgment of the author may be wrong. Mr. Robert Graves (I ought to say in passing that he is, so far as I know, the inventor of the method of analysis I was using here) has remarked that a poem might happen to survive which later critics called 'the best poem the age produced,' and yet there had been no question of publishing it in that age, and the author had supposed himself to have destroyed the manuscript. As I remember, one of the best-known short poems by Blake is actually crossed out by the author in the notebook which is the only source of it. This has no bearing on any 'conflict' theory; it is only part of the difficulty as to whether a poem is a noumenon or a phenomenon. Critics have long been allowed to say that a poem may be something inspired which meant more than the poet knew.

The topic seems to me important, and I hope I may be allowed to digress to illustrate it from painting. As I write there is a grand semi-government exhibition of the painter Constable in London, very ample, but starring only two big canvases, both described as 'studies.' Constable painted them only as the second of three stages in making an Academy picture, and neither could nor would ever have exhibited them. I do not know how they survived. They are being called by some critics (quite wrongly, I understand) the roots of the whole nineteenth-century development of painting. It seems obvious to many people now that they are much better than Constable's finished works, including the two that they are 'studies' for. However, of course, nobody pretends that they were an uprush of the primitive or in some psychological way 'not judged' by Constable. When he got an idea he would make a preliminary sketch on the spot, then follow his own bent in the studio (obviously very fast), and then settle down on another canvas to make a presentable picture out of the same theme. 'My picture is going well,' he remarks in a letter, 'I have got rid of most of my spottiness and kept in most of my freshness.' You could defend the judgment of Constable by saying that he betrayed his art to make a living, but this would be absurdly unjust to him; at least Constable would have resented it, and he does not seem to have had any gnawing conviction that the spottiest version was the best one. Of course, the present fashion for preferring it may be wrong too; the point I am trying to make is that this final 'judgment' is a thing which must be indefinitely postponed. Would Mr. James Smith say that the 'study', which is now more admired than the finished work, was a noumenon or a phenomenon? I do not see any way out of the dilemma which would leave the profound truths he was expressing much importance for a practical decision.[2]

The 'Mr. James Smith' in question is the author of a review of the first edition of Empson's book published in the *Criterion* for July 1931, which Empson quotes at length in the Preface to his second edition. As Smith sees it, Empson's method of analysis falls short of what Smith calls the 'main' business of criticism – to pass 'judgements of value'. Instead of deciding the proper *worth* of a literary text, according to Smith, Empson spends his time worrying over what nowadays might be called the indeterminacy of language, sometimes even venturing to speculate over 'conflicts supposed to have raged within the author when he wrote. Here, it seems to me, he has very probably left poetry completely behind.' Nor is Mr Smith entirely satisfied that Mr Empson has made sufficiently clear what he means by 'ambiguity':

> Is the ambiguity referred to that of life – is it a bundle of diverse forces, bound together only by their co-existence? Or is it that of a literary device – of the allusion, conceit, or pun, in one of their more or less conscious forms? If the first, Mr. Empson's thesis is wholly mistaken; for a poem is not a mere fragment of life; it is a fragment that has been detached, considered, and judged by a mind. A poem is a noumenon rather than a phenomenon. If the second, then at least we can say that Mr. Empson's thesis is exaggerated.[3]

Now there are clearly several differences between Smith's and Empson's approaches to criticism as recorded in these passages, but I want to pick up on a difference of a slightly different order from what might routinely count as noticeable: in Empson, 'judgment' is spelt without an 'e'; in Smith, with. I propose to call this difference a very interesting kind of *literary* detail, a term I use defiantly – for what else should it be called? No doubt Smith would refer to it as an insignificant preference or such like, and quite detached from the important matter of deciding the relative value of his versus Empson's approach to the task of criticism. For him it could represent only a 'phenomenal' under-standing of the text, which is to say a misunderstanding. But still I'm attracted to it, and puzzled by it. Could it be that Empson's 'judgment' marks a deliberate *lack* of 'judgement'? If so, this 'lack' can be signified only in writing – and *as* writing. Is this significant? It is, I think, if Empson's critical purpose is understood to be primarily *textual*; if what his project (similar perhaps to that of Chambers, as cited in the previous chapter) amounts to is an attempt at describing how texts 'work' rather than at evaluating what they 'mean'. For it would seem that the purpose of criticism as called for by Smith is nothing short of quite ridiculous. And surely the examples given by Empson, regardless of being 'true' or merely 'speculative', *prove* this to be the case. Why then does the idea persist that the critic's job is to arrive at 'judgements of value' on works of literature?

My initial answer is that it persists because it does not seem to carry the force of an *idea*, something that might be open to question and speculation, but rather of a *noumenon* (at least in so far as this concept seems to be used – or misused – by both Empson and Smith, albeit to slightly different effects), a thing-in-itself. Now I realize that the word 'idea' has several, as it were, both common and philosophical senses; it commonly refers to what is produced by mental activity, to some notion or other, to a fancy, an opinion or the gist of something. For Plato it meant something pure and archetypal, against which corresponding objects in the phenomenal realm exist as poor copies; for Kant it meant a product of pure transcendental reason; for Hegel, absolute truth; and so on. I am using it here in the common sense of a mental product, a notion of sorts, which would make it a problem to suppose that literary critics somehow just got it into their heads to pass judgements of value on works of literature. Given Smith's insistence on this being precisely what literary critics *should* do, however, his warrant must be the *noumenal* status of such activity, so that in passing judgement on a literary text the critic is engaged in the very thing-in-itself: namely, doing literary criticism. And the circularity of this argument (criticism is what criticism does; what criticism does is what it is) protects it from stubborn empirical instances to the contrary, such as Empson's examples of the Blake poem and the Constable paintings. But the point to notice here is that these examples raise a doubt about the status of the 'poem' *as a poem* and the 'paintings' *as paintings*.

In our paper 'Lowry's Envois', Alec McHoul and I tell the story of a supposedly 'lost' manuscript of a Malcolm Lowry 'novel' called *In Ballast to the White Sea*. There is actually some doubt as to whether the 'manuscript' ever existed as such, as a so-called phenomenon, which leads to the question of whether an instance of writing that was presumed to be at best only ever purely noumenal can be referred to as a 'novel'. Surely it has to have, or at some point have had, material form for this to happen (and I'm not suggesting that materiality is a *condition* of phenomenality). What would a novel that was purely noumenal (supposing this were possible) look like, in other words? Couldn't I

have written one of these already, carried it around with me in my head
for a time and then lost it somewhere, so that now I am able to refer to it
only as my 'former novel'? Indeed, Lowry himself employs this odd
locution in a letter to his agent Harold Matson, when he refers to the *In
Ballast* manuscript as having been 'lost' in a fire that nearly destroyed
his home.[4] Now, supposing that a manuscript did in fact at one time
exist, outside all references to it in Lowry's correspondence to and with
his agents and friends and so on – isn't there still some room for doubt in
calling it a 'novel', let alone a 'former' novel? Otherwise, couldn't I have
a 'novel' in the bottom drawer of my desk at this very moment, but only
in the form of an unpublished manuscript? Perhaps, more correctly,
such a manuscript would be in the ur-form of a novel, so that in order to
be fully realized as a *novel* it would have to be published (and therefore
institutionalized, being brought into relations with a *readership*). But
now what if I get my manuscript published and then live to regret it, as
happened to Hawthorne in later years concerning his first novel,
Fanshawe, which he published in his youth at his own expense? And so
the question remains: what status does the text that I might have
disclaimed now have? What would a novel look like that the novelist
had, in effect, crossed out? Or a poem that the poet had put under
erasure? Surely by crossing out the poem, Blake, in effect, cancelled its
'poem-ness'. If he'd *literally* erased it, by rubbing it out – or setting fire to
it – the 'poem' could have continued to exist only as a memory, in the
realm of the so-called purely noumenal. But not quite. For if such a
realm were possible then it would somehow have to be the absolute
other of the realm in which phenomena occur or exist. And this would
mean that phenomena exist outside or independently of any mediation
via the noumenal realm; in which case, how would anyone know that
phenomena were there, in the world, including us ourselves?

Take the case of the crossed-out Blake poem. Suppose this meant that
Blake did not want the poem to be published (it's possible it meant
something else: that he wanted it to be published under the crossing-
out, for example), or that the crossing-out indicated his decision to
refuse the poem's poem-ness (it's possible it meant something else: that
the crossing-out lent the poem its poem-ness, for example): who would
then decide that Blake was wrong, and on what grounds? First, let it be
noted that the crossing-out has been interpreted as a cancellation and
not as a poetic device. So clearly someone has decided what poetic
devices look like, and that this isn't one of them. The subsequent
decision to ignore the sign of the cross and pass an approving value-
judgement on the 'poem', or what remained of the poem after the

crossing-out had been erased, is foremost a practical one: its effect is to bolster the illusion of the noumenal status of those devices considered to be 'poetic', as opposed to signs belonging to other sets of writing practices. The crossed-out Blake 'poem' can *be* a poem, then, on the grounds only of its *possessing* some intrinsic quality, which is independent of the poet's judgement, knowledge or belief. But clearly it is not independent of *somebody's* judgement, etc. How could it 'be' a poem otherwise? Such a question has to do with what a poem which had no form outside of someone's head would look like. I am happy to admit that such a poem would have *some* form, inside someone's head, but what would this mean? For all intents and purposes, for all 'practical' purposes (to quote Empson), it would not *be* a poem. Now if the someone who 'wrote' it, while keeping 'it' inside their head, actually began to speak *about* the 'poem' to some others – then it would be very difficult to deny that the poem had become a (so-called) phenomenon in the world. At any rate, *something* would be in the world which had not previously existed: namely, talk about a poem, or about a thing so-called. And this would do; this would be enough to bring the poem to life – just as references to *In Ballast to the White Sea* are enough to satisfy the condition of that text being a novel.

This will no doubt seem a very strange 'condition', though. How could you *teach* this novel, for instance? My answer would be in the same way as the crossed-out Blake poem is taught without the crossing-out. Why not, after all? For surely if the Blake poem can be taught against the poet's wishes, as it were, then what a poem 'is' cannot be all that obvious. Apparently, indeed, the poem actually *exists* against the poet's wishes, so that one might well ask the question – who wrote this poem? It can't have been Blake, since he is the one who crossed it out, tried to get rid of it, pretend it didn't happen. So who, then? But of course the question seems preposterous. Of course William Blake wrote the poem. Then why did he cross it out? Having crossed it out, it seems to me, then the only way it can 'be' a Blake poem is for it to be read as it was written, *by the poet*: under erasure. Otherwise someone else wrote the poem and they ought to come clean.

Or – it actually doesn't matter to literary criticism who wrote the poem. Someone obviously did, and why not 'William Blake'? Poets are just a function of a certain discourse on a kind of writing which is understood to constitute 'poetry'. So as soon as a poem comes into the world, criticism actually doesn't need to know who made it do so: its only task is to evaluate the new arrival, check it for the worth of its poetic devices, and run a noumenal test on it. And whatever doesn't fit –

a crossing-out, an errant comma, a misspelt word, etc. – simply gets 'corrected' or else erased: not by the poet, but by the institution of criticism. So in fact all that is needed to do literary criticism is *text*, just text; the very idea that there are novels and poems already out there in the world prior to literary criticism *naming* them as such, simply denies the institutional force of what criticism does. *Criticism produces literature*; not the other way around.

My point is that, in fact, all that literary criticism has ever had to work with is text, and all that can be done with text is to show *how it works* – not what it means. The crossed-out Blake poem is a case in point: *it has been turned into literature from being text*. Paradoxically, this results in a loss of some meaning: at the very least, that the text is readable in, and as, its crossed-out form. But this meaning is crossed out, lost, once (the) text becomes (a) poem.

This means that what a poem 'is' (apparently) does not have anything to do with what it looks like, or with what might be called its phenomenality. If it were the case that a poem actually was a phenomenal event, then everything about its event-uality should be understood to be part of the poem itself. This would mean that poems should be published *as they were written*, since how they were written is actually what they are. And this is what I mean by 'text' – more or less everything.

For example: in the passage quoted from Empson's Preface, a comma appears outside the close quotation mark at 'study' in the penultimate sentence, which is very curious because Empson's practice elsewhere in his book is to put punctuation marks *inside* quotation marks, as happens four times in the passage above. These appear, in order, as follows:

- 'the best poem the age produced,'
- 'studies.'
- 'My picture is going well,'
- '. . . kept in most of my freshness.'

Routinely, then, Empson deals with the problem of punctuating quotations in the manner as shown. And, to some extent, it *is* a 'problem', given that one may choose to punctuate either inside or outside the close quotation mark, so long as the choice is kept consistent throughout the text (and generally confined, by convention, to commas and full stops).

So why the anomaly at ' "study",'? One answer is, of course, that this is a compositor's mistake, and that in fact Empson himself never made the decision to punctuate on one side or the other of imported (quoted) text: this was decided by the publisher – The Hogarth Press, in the case of the edition cited here. Indeed, writers never make this decision; it's always determined by the house style of a particular press, conforming usually to a regional style (UK publishers tend to punctuate outside quotation marks; US publishers inside). This means that literary criticism knows what to be critical of and what to regard as merely incidental, mere phenomena – more or less any inconsistency. But on what grounds? Precisely those enabling, or seeming to enable, a distinction between the noumenal (literature) and the phenomenal (writing). On these grounds, the anomaly at and of ' "study", ' in Empson's Preface does not count as a critical fact about the Preface 'itself', any more than the crossing-out counts as a fact of Blake's poem. And this is weirdly contradictory. For whereas on the one hand the agent of the incidental phenomenon is the poet himself (who crossed out his own 'poem'), the agent of the incidental phenomenon in Empson is not the critic himself but his compositor or publisher. In each case the phenomenon is not of proper concern to literary criticism. Clearly, in the case of the crossed-out Blake poem the crossing-out has been overlooked as incidental; and it would be difficult to imagine a critical account of Empson's book (with the present exception) which took any notice of the phenomenal difference at and of the comma outside ' "study" '. So what counts as 'incidental' cannot be simply whatever is decided that a text's producer didn't do, or can't be held accountable for – such as a misspelt word, a wayward punctuation mark or error, a typographical inconsistency – because sometimes the agent of the 'incident' is in fact the one who 'wrote' the text, even if he crossed it out.

Contrary to this being a mere quibble, moreover, it cuts at the very heart of literary criticism. For literary criticism to perform itself, *texts must have agents*. This does not have to be the case (actually, it isn't), but certainly it is virtually so. Otherwise, as even Empson puts it, criticism lacks any sense of its own purpose unless critics 'put up some pretence of understanding the feelings of the author in hand'. What an unauthored text would look like (and I don't mean an anonymous text) is anyone's guess as far as literary criticism goes, which could grasp this idea only as an oxymoron – like a square triangle or a carnivorous cow. So the pretence is in fact an *imperative* of criticism, and something very real. But that doesn't make it any less of a pretence, as at least Empson was astute enough to infer in 1947. Somewhat later and elsewhere, Foucault gave the pretence a name that captured its imperative status: he called it the author function.[5] And it is this which is at work in any 'critical' decision to disregard the comma outside 'study' in Empson or the crossing-out over (not *of*) the Blake poem. These phenomena are understood to be 'unauthored', in other words: simply a typographical mistake in the case of Empson; in Blake's case, an inexplicable lapse of judgement. In neither case does the phenomenon constitute an *organic* component of the text itself.

And this is surely very interesting. What it must mean is that the author function works to sort out the essential from the incidental 'in' any text, as well as between texts (novels are essential to any idea of literature; shopping lists, even if written by novelists, are incidental). But this is just another way of purporting that whatever is 'essential' is in fact *self-evidently* so, and likewise whatever is 'incidental'. So the author function comes to serve as the justification of whatever factors criticism 'decides' retrospectively are, self-evidently, organic components of a text. Therefore William Empson is not the author of the typographical mistake that occurs at ' "study",' in his text, and so this event or phenomenon, although clearly actual, is not actually a part *of* the text; it's a printing error, part of the book. And this would still be so even if it were found that the mistake occurs in Empson's own manuscript: it would still be a *writing* phenomenon belonging to a material process, that of the *book's* production, and not part of *the text itself*.

Now it should be clear that I do not accept this distinction between the noumenal and the phenomenal, the text and the book, literature and writing. After Derrida, I can't. But nor does it seem possible to do so after Empson, who comes before Derrida, although admittedly I'm reading him in Derrida's wake. (And neither is this necessarily wilful or capricious, given that the number 'seven' figures so conspicuously, if that is the right word, both in Empson's account of ambiguity and in Derrida's of the postal metaphor in his *Post Card*; while another coincidence of this kind is discussed in 'Lowry's Envois'.) For surely the examples that Empson gives, of the crossed-out Blake 'poem' and the Constable 'paintings' that were never intended for exhibition, serve to show that the division between text and artwork does not at all describe a difference between noumenal and phenomenal states of being. The division is an after-effect imposed on writing (as the name of a material process); it's an institutional effect, a determination of the practices of criticism.

It might be noticed from the previous paragraph that my use of the word 'text' has shifted: from being on the side of the 'noumenal' and 'literature' to become the opposite of these. I had better explain. I'm trying to find a word that describes what literary criticism practises itself on; what is 'it' that literary criticism is (the) criticism *of*? The word is clearly not 'book', or 'books'. *Jane Eyre* is not *a book*; novels are not books; and *Hamlet* is certainly not a book – nor is a Blake poem, or Melville's 'Bartleby'. 'Book-ishness', or the condition of being *in print*, is simply the *precondition* of 'literariness'; at any rate it's the precondition of a readership for literature, and literary criticism is not renowned for an interest in questions concerning readerships. Literary criticism cannot be confined to a study of what the publishing industry produces, commodities in the form of books.

Instead, what literary criticism studies is *literature*. Nothing would seem more obvious, almost tautological, once written down like that. But still it begs the question: what is literature? I have said that it isn't books – these are just a technology for 'doing' reading. And the only other word I can think of is 'text': *literature is text*. What I mean by this is that literature is something; it takes place in something and as something, and somewhere. And if literature is indeed something that has been achieved through a critical act or a judgement of value, then that judgement has not been conferred on a *book*. Nor has that judgement occurred in a *geophysical* 'somewhere', but rather in what I want to call for now an institutional space or in the space of a community. Again that is what I mean by 'text', although I do not have a neat definition of what this means. That is the point: the thing-in-itself (as noumenon) always escapes definition, any effort to enframe it, to close off what it means from what it doesn't. Because I certainly do not think that the text *doesn't* mean the book. I do not think that I am *not* reading *Jane Eyre* when I'm reading *Jane Eyre*, which I always do with a book in my hand, or at any rate with one open in front of me. (Even if I were to do this 'in my head', I don't know that I could do so without thinking of the words *on the page*.) So there is actually a point to my having shifted the sense of 'text' from being in opposition to 'book' and then to 'artwork' in a single paragraph. I wanted to establish that literary criticism does not study books, the commodified products of a publishing industry; what it studies is *text*. But I also wanted to say that the effect of a critical industry is to turn texts into works of art (works of literature in the case of a 'literary' critical industry).

On the one hand, then, I do not have a word other than 'text' for what literary criticism studies. On the other, I know that this is not the word that literary criticism uses to describe what it studies, that word being 'literature'. But this simply cannot be the case. Why? And I return to Empson, who cites the specific instances of the crossed-out Blake *text* that became a poem, and the Constable *studies* that became paintings (which, for some critics, constitute nothing less indeed than 'the roots of the whole nineteenth-century development of painting'). Now how can this be? William Blake sits down to write a poem, thinks to have done so, then changes his mind and crosses it out. Or: William Blake sits down to write a poem; he writes words, then he crosses them out, then he thinks to have done so. And then literary criticism decides that Blake has in fact written a poem after all, but that the poem is in fact *only the words*, ignoring the crossing-out. John Constable has an idea for a painting, so he makes a study of it; he does a quick sketch, adds a few more touches and takes it to a second stage of near completion, but he never finishes the thing. He has another idea, makes a study, does a sketch, adds touches, gets this one to the second stage as well. In neither case does he ever intend to exhibit the *unfinished work*, which he never thinks of as a painting. Even if he'd wanted to exhibit them, the London Academy at that time would never have allowed it: they were one stage short of being paintings – they were *studies*; more or less officially, they were just text. And then, over a century later, art criticism decides that these studies are Constable's most important *paintings*.

The question is: why are Constable's *studies* not just (so-called) phenomena? Why aren't they just text, like a shopping list? On what basis have they been transformed into (so-called) noumena, into *paintings*? On what basis has the text-ness of these (incidental) things-in-themselves been transformed into something else, something other and transcendental, something *more* (and less) than text? And why, if indeed Blake did mean the crossing-out to signify the failure of his words, in his opinion, to achieve poemness (that they were no more than just text, like a shopping list), did his words become a poem; or why, if he meant the crossing-out to be part of the poem, have only the *words* achieved poemness?

It's not that I think there are simple answers to these questions. In this regard I think that Empson is right: the final judgement has to – *must* – be 'indefinitely postponed'; and I want to try to show why here, by trying to show how. I want to try to show that what he calls 'the dilemma' of deciding whether a poem, or any work of literature, is in fact a noumenon or a phenomenon, is exactly that – what must always remain undecidable. Undecidable, but not in any (what he calls) 'practical' sense. This is not to say that practical considerations do not have any bearing on the productions and circulations of texts. Nor is it to suggest that even everyday pleasures and personal passions do not intrude on other domains of a text producer's life, including the process of writing this book. But I cannot give an account of how or why they intrude, any more than I can explain the 'accident' of the seemingly misplaced comma outside of 'study' in William Empson's Preface. And

I certainly don't want to go looking in my 'unconscious', or in his, for an explanation of why I like certain things or why there is a wayward comma in *Seven Types of Ambiguity*. Because if meanings can be sourced to the unconscious, then the theory ought to be able to give at least an in-principle explanation for the 'typographical' anomaly in Empson. But the problem is that it wouldn't then still be a 'typo' (in fact it would have to have been always already not this), and yet it seems rather far-fetched to call it a 'psychic' event. Now I'm not saying that a psychoanalytic account of the comma would promise to explain its textual appearance in terms of a history of the psyche, but that is what it would be committed to in principle. And if the principle of the Freudian slip extends even into the domain of typography, the unconscious being the agent of all sign-use, including 'mistaken' instances, then surely nothing can escape this principle. If nothing escapes it, isn't the principle itself subject *to* itself? If 'conscious' sign-use is always underpinned by desires and motives hidden in the user's unconscious, so that an 'intended' effect is always just a slip away from being 'unintended', and if this is *a principle of intention*, from which not even psychoanalysis is exempt – how could psychoanalysis ever be in a position to know an unconscious effect when it saw one? For such an effect would always be 'there' in principle, at every exchange of the sign, on every occasion of its use, including every *psychoanalytic* occasion and exchange.

So no, I don't think Empson's comma is a Freudian slip. I do not even think the comma is Empson's, nor that it is in 'error'. It isn't *Empson's* comma and it's not a mistake, because for it to be 'his' it would have to be distinguishable from what was not his (and I don't know the grounds on which this distinction could be made); and for it to be in 'error' it would have to be distinguishable from what was otherwise 'correct'. But the point is that for something to be 'correct' it has to be, in principle, capable of being in 'error'. This is not to return to a theory of the unconscious; it is a statement of the structural necessity of the possibility of error to every exchange and occasion of the sign. Following Derrida, this necessity can be called *writing*, as the name of a technology conditioned by a lack and structured in difference. But this condition also structures speech, traditionally the pure form of sign-use from which writing in the West is always understood to lapse. However, as Derrida writes in *Of Grammatology*:

A signifier is from the very beginning the possibility of its own repetition, of its own image or resemblance. It is the condition of its ideality, what identifies it as signifier, and makes it function as such, relating it to a signified which, for the same reasons, could never be a 'unique and singular reality.' From the moment that the sign appears, that is to say from the very beginning, there is no chance of encountering anywhere the purity of 'reality,' 'unicity,' 'singularity.' So by what right can it be supposed that speech could have, 'in antiquity,' before the birth of Chinese writing, the sense and value that we know in the West? Why would speech in China have had to be 'eclipsed' by writing?[6]

On every occasion and at every exchange, then, the sign is conditioned by this principle of the signifier's iterability, which must also structure the signified in difference and *as* difference. This must be and always has to have been the case, so that there's no question of there having once been an 'original' or 'pure' sign from which any 'subsequent' exchange or occasion could be said to mark a lapse or to be constituted by a lack, or to be in error. For the sign is always already lapsed and lacking, and no less in its spoken than its written form; the sign is always 'in' error.

Such an instance, indeed, occurs 'in' the very passage just quoted (or is performed 'by' the passage, or 'as' the passage). Turning to page 334 of Spivak's translation, it can be found at note 43 that Derrida's quotation marks refer to a paper by Jacques Gernet, whose title Spivak gives in French: 'La Chine, Aspects et fonctions psychologiques de l'écritures'. Translated, this might read as 'China: Aspects and Psychological Functions of its Writings'. But this is not the title that appears in *De la Grammatologie*, or 'in' Derrida as it were, where on page 138 at note 43 the title is given as *La Chine, Aspects et fonctions psychologiques de l'écriture*. Despite Derrida's italics (which I'm happy to attribute to a non-Anglo convention), this is still the title of a paper, not a book; so that in English it might read as 'China: Aspects and Psychological Functions of its Writing'. In other words, Spivak has added an 's' ('l'écritures' rather than 'l'écriture'; 'writings' instead of 'writing') – at the very least, someone has. Or better still: an 's' has found its way into the *Grammatology*.

Another mistake? But surely this itself would be a mistake, since we've already found mistakes to be of at least two separate orders. There is the kind of mistake that occurs in Empson, where 'his' wayward comma and the extra 's' supplied by 'Spivak' are of the order of a typographical error. (This is to ignore for now that Spivak's 's' is ungrammatical, inasmuch as a correct plural would be *les écritures*. Before the question of responsibility arises, then, there is the question of what the mistake actually comprises.) Then there is the kind of mistake that *would* be made if it were supposed that the crossed-out Blake poem is not in fact *a poem* or the Constable studies are not in fact *paintings* – an error of judgement. In order to value the Blake and Constable texts as works of art, however, it must be able to be decided when something is not a mistake. And this something cannot be called an 'intention', since it is precisely Blake and Constable's intentions that must be disregarded in order to convert their texts into a poem and two paintings. So if the first move of literary criticism in the case of the Blake text is to overlook the poet's intentions in order to constitute the text as a poem, and if this move is made of necessity, surely it can be seen that this is precisely the first move of literary criticism with respect to *every* text. The Blake case is not a special instance, but rather a general condition, as Empson did not miss understanding in 1947: citing Robert Graves, he remarks that 'a poem might happen to survive which later critics called "the best poem the age produced," and yet there had been no question of publishing it in that age, and the author had supposed himself to have destroyed the manuscript'. He then repeats the point in his own words two sentences later: '[c]ritics have long been allowed to say that a poem may be something inspired which meant more than the poet knew.'

So in fact criticism has never been terribly concerned with authors' opinions of their own writing. Authors are important only as agents of text production (someone has to write the stuff), but it's criticism that decides which instances of writing count as literature. Nevertheless, having made its decision, criticism then finds that authors are extremely useful in their *function* as the agent of textual meaning and source of exegetical felicity, so that interpretations of a literary work are always framed around and by a notion of its author's intentions, taking into account the limits of semantic possibility for a given age and so on. Yet there seems no question that the text survives the operations of this *pretence* to remain what it has always been: the scene of writing, the subject of criticism.

In this dis-guise, the text refuses both the *belonging* and the *designation* of a mistake. Mistakes do not belong anywhere, and they do not belong to anyone. Where would the rightful or the proper place of a mistake be? Who would be said to own it? To this it might be objected that it depends on the kind of mistake. So, for example, if I had written above that Nathaniel Hawthorn is the author of 'Bartleby', this might be taken to constitute both *my* mistake and just *a* mistake. It would be my mistake not to have attributed authorship of 'Bartleby' to anyone other than Herman Melville, but it could be just a mistake to have misspelt

Hawthorne's name. While it is true that the misspelling would be understood to have an agent (me), and in this sense to be 'my' mistake also, nonetheless I would not be its *author* in the sense that Melville is understood to be the author of 'Bartleby'. And certainly if Melville's name appeared without a second 'e' on just one occasion in this chapter, this would not be said to be my mistake at all but just 'a' mistake, a typographical error. There would seem to be three categories of mistake, then: mistakes with authors; mistakes with agents; and what might be called mistakes-in-themselves. This last category, being the purest kind of mistake, could therefore be described as purely textual, and it is for this reason that I think it is the most interesting – despite and because of its being of least interest to literary criticism.

Mistakes-in-themselves, errors of a purely textual kind, are not gener-ally countenanced by literary criticism. These are the provenance of literary historians, but not hermeneuts. Criticism proper, as it were, occupies itself with exegetical accounts of what texts mean, and it always undertakes to do so on the basis of what might be called a 'pragmatics' of reading. This is to suppose in other words that texts *have* a kind of 'univocity', a sort of noumenal truth or 'true worth' which is conveyed by but not exactly contained in the materiality of writing, the physical phenomena of signs (see Chapter 7). When the pragmatic purpose of reading is to get at this univocity, this noumenal truth, then it's easy to discount some signs as being out of place, as not actually belonging to the text although they might in fact be 'in' the text. So how did they *get* 'in' there, and where do they belong? How did an extra 's' find its way into the *Grammatology*?

Let's not play with the idea that Spivak put it there intentionally in order that I might find it and write about it here. Not yet, anyway.[7] And nor does it have to be supposed that a compositor put it there deliberately as a practical joke, or to make Spivak look careless. No one *wrote* it, in other words; which leaves only the possibility that the 's' wrote itself, slipped into the *Grammatology* of its own accord. But as far-fetched as this might seem at first, it is in fact no more than a variation on a commonplace of literary criticism, and one that Empson has already been quoted to recount: '[c]ritics have long been allowed to say that a poem may be something inspired which meant more than the poet knew.' It is an article of faith, then, that there are things 'in' texts which are not consciously put there by their agents of production – these things are called meanings. And if that is what they're called, then

I want 'mistakes' to be meaningful too. No less than metaphors, I want *mistakes* to 'have' meaning.

All this would entail is a rethinking of what meaning means, along the lines of dissociating a meaning from 'its' producer. The examples of the crossed-out Blake 'poem' and the Constable 'paintings', I think, are proof that criticism already thinks this way, but without necessarily knowing (or admitting) it. From the very start, all that is required to do literary criticism is text; not a theory of authorial intention, nor a deliberate valorization of speech. Text on its own will do. So why stop with a pragmatics of the *partial* text? Why separate the text into instances of 'itself' (themata) and 'not-itself' (errata), consigning the latter to a place *outside* 'the' text?

The word 'theme', deriving from two Greek words, literally means 'to place a proposition'. So the theme of a text is a proposition that has been put in place, and the obvious assumption is that it has been put there by the author. The word 'erratum', from the Latin, means 'to stray' and usually refers to a printing error that has been corrected, brought into line, the intended form having been put back in its place. So aren't what I have been calling 'errors' (the aberrant comma in Empson, the extra 's' in Spivak's Derrida) more commonly known as *errata*? Haven't I effectively turned them into such, at least, by remarking on them here? For surely in the forms in which they appear in *Of Ambiguity* and *Of Grammatology* it is quite proper to refer to them as being outside their texts, since they are obviously astray; they don't belong there. Nevertheless, there they are. And this fact can be got around only if the text is understood to mean not a concrete phenomenon, but a noumenal abstraction; only if, in other words, the idea of the text is text itself. According to this idea, the place of the theme (the space of meaning) is determined and secured *in advance of any critical engagement with the text*; otherwise the holders of this idea would have to believe that they are in fact the producers of textual meaning, that the theme is put in place by the text of criticism. Since it is clearly not authors who do this placing (Blake, after all, crossed out his poem; and '[c]ritics have long been allowed to say that a poem may be something inspired which meant more than the poet knew'), it must be performed by *the author function*, this function acting as a licence to attribute meanings to the text of the author.

So in what sense, then, is an erratum any less in place than (say) an irony, or any other critically approved in-stance of a literary device? But perhaps the question should be: in what sense is an irony, a figure, a trope *in place* at all? And this in turn may be to put the question to literary studies, as a *teaching* discipline, as to what it is a discipline of teaching *for*. Or perhaps this is not the question; it isn't that literature should be 'for' *something*, but that it can be used for many things. It *can* be used for moral instruction, 'self-development', inculcating national-ism and so on; or even for the admirably practical purpose of preparing future citizens to write not 'imaginatively', but well. Nor would the set of works comprising the category 'literature' necessarily have to be the same for each of these uses: students could just as easily be taught practical writing skills, as opposed to 'appreciationism', by reading outside the canon – they could as easily learn to write 'well' by reading William Gibson as Jane Austen, if this were the desired end of a literary studies curriculum. But it isn't, certainly not in high schools and university undergraduate programmes where the emphasis remains on educating, and (re)producing, the liberal individual. (It's true that being able to write well is one of the very hallmarks of this subject, but what 'well' means is confined generally to notions of plainness, politeness and self-effacement, and understood to be a positive value in itself. The distinction is fine, but I'm thinking of being able to 'write well' in terms of a purely social utility – as a kind of power tool rather than a 'moral' accomplishment.) For as long as this continues to be so, it seems to me therefore that what goes on within the literature department cannot hope to understand the world or begin to change it.

All of a sudden it looks like what I'm saying is that I wrote this chapter to change the world. While I can't bear to read back over the sentence I have just written, I know that I would admire the desire it harbours if someone else had written it, even if at the same time I were to scoff at their ambition. So I'm going to pretend that I did not write the sentence, but that it just found its way in here as did the 's' in Spivak's Derrida. Under this pretence, then, the sentence becomes text, in the same way that it would become 'intentional' or 'authored' under the pretence that I actually wrote it. In other words, its 'authoredness' is not a condition of its making sense, this being an effect of its *textuality*. It is *as writing* that the sentence means.

And it is precisely in this capacity that the text is lacking, from the point of view of judging literature. The primary function of literary criticism is to supplement this 'lack' with a purpose, in order that 'mistakes' (especially of the third variety) can be overlooked. In effect, the idea is to *bring* the text to mystery, to produce it as an enigma, a unified *work* of the imagination. Once brought to this condition, the work is then able to be divided into its so-called noumenal and phenomenal parts. These are able to be designated by several terms:

work and nonwork, work and text, text and outside-text. The second term of each pair is itself able to be divided in turn into what might be called organic and nonorganic parts, or necessary and accidental phenomena. So a particular word is necessary, though the typeface in which it appears usually is not; or the setting of a poem is organic but the volume of space surrounding it usually is not; or it is necessary that an apostrophe of omission should appear in the word 'it's' for 'it is', although it would be an accident if it didn't. But there is never any such instance that actually belongs to the work (or the text, depending on the lexicon) itself, *as noumenon*. The work itself (its true worth), in other words, is irreducible to mere phenomena, although these provide the only means of access to it.

Nothing could be more straightforwardly unremarkable, then, than the undecidable nature of the literary work. Whether its undecidability is couched as ambiguity or as adestination (among many other possible candidates), such terms go to show only that literary studies cannot rely on a hard and fast distinction between proper and erroneous examples of literature. For it is the very category of the literary itself that must remain unsettled and unsettling, and about which, as Empson put it – long before Derrida – there is nothing final to be said and everything left to say.

This is not however to assert that literature is an inherently special or privileged mode of textuality. As we will see in the following chapter, for all that literature may be understood as a site of undecidability – and for all that this ensures its irreducible singularity *as* such a site – it would be a mistake to regard undecidability as an exclusively literary condition rather than a condition of textuality in general.

9. Gilligans Wake

I half remember an old joke from *Gilligan's Island*. At the end of an episode in which – I could be wrong about this – the castaways find out they've been eating radioactive vegetables (saved by the Professor, of course, who dashes off a cure in true bricoleur style out of something very ordinary to hand – soap, I think), Gilligan is seen reading a book (previously washed ashore, no doubt, like every other *deus ex machina* in the series) whose title is *How to Tell a Mushroom from a Toadstool* by the late Dr So-and-So. No – I'm sure that's the wrong episode, but in any case that is the joke. Not very well told, perhaps, but there you have it . . .

Soap

So: we already know about soap and its distinction from chlorinated cleaning agents. As Barthes writes:

> The relations between the evil and the cure, between dirt and a given product, are very different in each case.[1]

After Barthes, soap became a figure of conceptual apparatus. Remotivated, it then became endlessly refigurable.

And now, following Gregory Ulmer's *Teletheory*, it figures here as a 'research pun', a 'puncept', in a series of such figures (*inter alia*, 'mushroom') at work in the present text.[2] But also worked on *by* the present text, to be set in motion in a context even where it should be most at rest. By chance, in other words, I find that 'soap' is immediately preceded in my dictionary by the entry 'so-and-so', thus linking the washed-up Professor with the author of the washed-up book in my anecdote.

Although not by chance, surely. For surely it must be necessary that I or anyone consulting *The Random House College English Dictionary* will find 'soap' so listed, under 'so-and-so' – and, for that matter, followed by 'soar'. Just as the word 'mushroom' is listed in most dictionaries followed by 'music'.

It is (or sadly, was) John Cage who reminds us of this, the lexicographical place of 'mushroom', where the word is in little danger of being mistaken for the wrong thing. But out in the wild, as it were, where the signs are not so easy to read, mistakes can lead to death. Out there, the difference between poison and cure is no laughing matter; it's nothing to poke fun at or pun on or quibble over at all. If it is a game, then it's a deadly serious one, the mortal effects of whose miscalculation can be safely forecast.

So much for wild mushrooms. But what of the other kind, found in books? This

> may be understood as a model mounted in a discourse for allegorical purposes. Indeed, the mushroom turns out to be a good emblem for what Derrida calls the 'pharmakon' – a potion or medicine that is at once elixir and poison (borrowed from Plato), modeling what Derrida calls (by analogy) 'undecidables' (directed against all conceptual, classifying systems). The undecidables are:
>
>> unities of simulacra, 'false' verbal properties (nominal or semantic) that can no longer be included within philosophical (binary) opposition, but which, however, inhabit philosophical opposition, resisting and disorganizing it, *without ever* constituting a third term, without ever leaving room for a solution in the form of speculative dialectics (the *pharmakon* is neither remedy nor poison, neither good nor evil, neither the inside nor the outside, neither speech nor writing.[3]

And this is not a mistake, not here in any case. For Ulmer never closes the parenthetical citation from Derrida which he opens at '(the *pharmakon*', although a closed parenthesis does appear at 'writing)' on this page in Derrida's *Positions* itself ...

Cite

We might therefore ask what the criteria are for the selection of quoted material in literary (here, Pynchon) studies? Traditionally they are probably based around a notion of support for an argument or thesis. But since the thesis is woven out of a selective reading of the text anyway, the selection of the quotes and the thesis they support must be mutually constitutive.[4]

So, perhaps, one way to begin to learn to think to write with undecid-ables is to understand the problem of citing not simply as a writing problem, but *as* the problem of writing – that this is how we must learn to write in the post-age, the era of telecommunication. For example, following (or skewing) a practice used by McHoul and Wills in their chapter 'Telegrammatology', we might choose to cite ourselves accord-ing to what may appear at first to be a strange principle. The principle requires that we summarize the story so far (I hesitate to call it an argument, at least not yet) by quoting the first sentence of each of our own paragraphs above. Thus:

> I half remember an old joke from *Gilligan's Island*. So: we already know about soap and its distinction from chlorinated cleaning agents. And now, following Gregory Ulmer's *Teletheory*, it figures here as a 'research pun', a 'puncept', in a series of such figures (*inter alia*, 'mushroom') at work in the present text. Although not by chance, surely. It is (or sadly, was) John Cage who reminds us of this, the lexicographical place of 'mushroom', where the word is in little danger of being mistaken for the wrong thing. So much for wild mushrooms. So, perhaps, one way to begin to learn to think to write with undecidables is to understand the problem of citing not simply as a writing problem, but *as* the problem of writing – that this is how we must learn to write in the post-age, the era of telecommunication.

Nothing would seem less open to question than what we have just done, except that we might want to put the teleological question: why? But is it really as innocent as all that? I can think of at least two questions to ask at this point, which relate purely to the practice and not the purpose of our exercise: why did I *indent* my own words, and why didn't I quote the first sentence, albeit from McHoul and Wills, that begins the present section? There is at least some room for debate here, in so far as by indenting my own words I have lent them the status of being imported from elsewhere, in accord with the standard conven-tion, which is at the same time presumably the reason I did not include the sentence from *Writing Pynchon* – that it does not belong here in the present text except under strict conditions which identify it as being on loan and not to be confused with stolen property. But now what exactly is it that I have borrowed from McHoul and Wills? It cannot be their *intention* that I've borrowed (not after all these years), but neither can it be, straightforwardly, their *writing* – not if in principle I could have 'quoted' them, as Ulmer 'quoted' Derrida, but forgot to include all the punctuation marks (as I could have done). So what I have quoted is in fact the *meaning* of the passage, which is in fact its semiosic potential. Now if we accept that this potential is already critical and dynamic in what we can allow to be the passage's 'original' context, the book *Writing Pynchon*, such that what it means is never fully determinate and univocal, then we must allow that what it means in another context (here, say) either exceeds its polysemy at source or at least is never knowably coincidental with it. In other words, what it means *in other words* is something different or else unknowably the same, and there-fore different. But of course it was (is) already different from the start; so that in fact there is no question of its being identical (here) with itself (there, in *Writing Pynchon*), either materially or semiotically. And if the passage is never self-identical – then, when I quoted it, what exactly did I quote?

Presumably, we must have felt that something came across with the passage in its transfer to the present text – that something arrived in the crossover. This at any rate is what we would have felt had we been working with the *letter* as our correspondence-model. But what if our model of correspondence were (dash) the telegram (dot dot dot)?

> telegrams are not sorted, at least not in the same way as letters. They have direct delivery, though subject still, of course, to any number of relays and delays. Whoever pays the price of the telegram does so by the word and so may well pay the price, also, of truncation and lost coherence. On this model, suppose we were to take our text-bits first with no extra- or intra-textual semantic destinations 'in mind'. This would mean taking the text as, primarily, a material signifier. For example, we could cut out, collect and send, say, those parts of a text which held the same material, structural or formal positions. This would constitute a new form of 'identification' with the text.[5]

And this, taking the common structural position of a set of data as our organizing principle, is what we did when we made up an identikit version of the (then) present text out of the first sentence in each paragraph. But not even such an apparently 'innocent' organizing principle is without a certain agony, arising from the difficulty of determining the constitution of the 'same'. So it is with the telegram, too, of course: despite its reputation as a model of error-free communication, 'the telegram is liable to be *erratic*'.[6] As an error-*ridden* means of sending a distant message, the telegram may thus 'mean the excursus or projection on writing (*tele*, distant; *gramma*, that which is written)',[7] sending itself, perhaps, as a figure for change, in the form of a paradigm shift from a concern with the teleological towards the telegrammatological. According to this (new) paradigm, critical/writing is in fact always already a writing of – and as – telecommunication, which we can no longer miss understanding as a form of writing-in-general rather than a special instance of communication, or writing-as-such.[8]

Sine

Somewhere, already, this isn't new(s). Playing in the background as I write, a record by Grand Master Flash & the Furious Five pumps underneath my thoughts. In the mix I can hear the sampled sounds of other records, by other artists, re-sined by the Master and his DJs ...

Is this the space of hip-hop and high theory, where work becomes text?

> So the Text: it can be it only in its difference (which does not mean its individuality), its reading is semelfactive (this rendering illusory any inductive-deductive science of texts – no 'grammar' of the text) and nevertheless woven entirely with citations, references, echoes, cultural languages (what language is not?), antecedent or contemporary, which cut across it through and through in a vast stereophony. The intertextual in which every text is held, it itself being the text-between of another text, is not to be confused with some origin of the text: to try to find the 'sources', the 'influences' of a work, is to fall in with the myth of filiation; the citations which go to make up a text are anonymous, untraceable, and yet *already read*: they are quotations without inverted commas.[9]

But how to write this text, or rather how to write it with a positive value since it is already (being) written? How to quote without quotation marks in the era of telecommunication, the age of television? By means, surely, of one of the latest writing instruments – the synthesizer, with its capacity to sine again. By such means cultural history may be sampled, and a synthesis produced between 'antecedent and contemporary' practices, popular and academic discourses, personal and institutional forms of knowing. A synthesis moreover that might mark a new set of relations between the academic and the popular, where the former's presumption to teach is remotivated into a desire to learn from the latter.

At the same time, this set of relations might not be characterized as something fundamentally 'new' so much as constituting a new *instauration*, a discovery process which is also – and this may be in the nature of discoveries (think of the ongoing 'revival' of 1960s clothes and music fashion, which is also happening for the 'first' time) – a process of recovery as well. For the apparatus of academic pedagogy, this might mean a return to 'alternative' methods of analysis and understanding. But first we need to understand that the history of method *'has* a history' the privileged medium of which is alphabetic writing producing a form of analytico-referential discourse ordered on the model of the sentence, 'the syntactic order of language'.[10] Good syntax (good science, good semiotics) in the form of the well-ordered sentence thus provides for a clear understanding of concepts, which in turn gives access to the world of objects. But what if this clear-sighted telescopics were to be remotivated in terms of a soapy teletheoretics: what forms of wide scope would then be possible with so many suds on the lens?

Suds

'Suds', like 'scissors', is always plural. So here, in our discourse on method, we may think of suds as a form of cut-up, produced by a new instrument of writing – the telesoap. On television, soap is the lowliest form of the lowliest medium's textual production, the epitome of academic fear and loathing of the popular – the absolute other of alphabetic writing celebrated as the well-turned phrase, the properly syntactic sentence, the academic essay. Telesoap is dirty, vulgar, cheap; its production values trashy, its sentimental values false.[11] Above all it patterns itself after the reel world, failing in its resemblance to the real.

Paradoxically, the general academic disregard for television, and the disrespect for telesoap in particular, arises from a 'lack' that TV shares with *writing* – distancing effects that threaten to increase the risk of error whenever messages are sent (see Chapter 8). Pedagogy has learned therefore to trust only analytico-referential discourse as the form of writing best equipped to control these effects, making it possible for sentences to be revised until they (re)produce a desired sense. But this is not what always happens:

> The discourse of the telegram ... has one notorious feature. While it is primed for accuracy and economy, as we have noted, what is overlooked is that these two aspects can cancel each other out; the telegram is liable to be *erratic*.[12]

Indeed it is, as this performance shows. For how could something be 'notorious' *and* be 'overlooked' at the same time? But if this is a demonstration of error, and precisely of the order being argued, can it be called a *desired* effect of the discourse? Either way, however, it is still what Barthes might call a *punctum*, which the authors themselves gloss as 'a point to be noticed, a point which ... sticks out'.[13] In this way the 'error' punctuates the text after a fashion no less essential to the writing than a comma. So we might call the 'error' a *telegrammatological punctuation mark*, in order both to distinguish it from and conflate it with the set of standard (grammatical) signs of punctuation with which we are familiar and without which we could not write at all. By the same token, we must not miss understanding that we could not write without these *tele*grammatological punctuations, the erratic effects of writing-in-general, although it's certainly true that we use (can only use) analytico-referential forms of writing as if we did. But while writing *as if* we were able to control the text, however, unbeknown to us the text is always writing itself (and 'us'); for, as Barthes remarks, the *punctum* does not designate only a wound, but 'also a cast of the dice'.[14]

It is these chance effects of writing, leading as they do to the ultimate possibility – nuclear disaster – that concern, or should concern, the critical strategist in this age of utter uncertainty. For the apocalypse – as the perfect reading, the moment of the letter's arrival, the revelation scene or final destination – has not yet come about, which is why, precisely, it is 'a structure common to every trace' and therefore is 'too important a thing to be left in the hands of a fringe of experts or lunatics'.[15] Too important, because the nuclear question haunts all instances of writing-as-such, so that – and here I quote McHoul and Wills quoting Derrida[16] – the question of *strategy* takes on a new and urgent significance:

We are representatives of humanity and of the incompetent

humanities which have to think through as rigorously as possible the problem of competence, given that the stakes of the nuclear question are those of humanity, of the humanities ... In our techno-scientific-militaro-diplomatic incompetence, we may consider ourselves, however, as competent as others to deal with a phenomenon whose essential feature is that of being *fabulously textual* ... In the undecidable and at the moment of a decision that has no common ground with any other, we have to reinvent invention or conceive of another 'pragmatics'.[17]

One such name, aside from the general nomenclature of 'nuclear criticism', for this new 'pragmatics' or remotivated 'invention', is what Gregory Ulmer calls *teletheory*. At once the principal aim and the grounds of the new pragmatics, although not contrary to the strategies of heuresis (hermeneutics) which have come to occupy the centre of humanities work, is *euretics*:

> There is no need to be against hermeneutics in order to be for euretics, only that euretics provides an alternative to interpretation that has been lacking in most of the discussions of the problem. Hemeneutics, in any case, comes after euretics, applied to the invention as if it came from another, as the discourse of the other, to see what has been made; to note its meaning, value, or beauty.[18]

The new instauration of euretics (*inventio* and or as innovation), then, seeks to break down the distinction between criticism and writing, striving for a critical/writing practice that is both 'theoretical' and 'performative' at the same time: in a word (but one I hesitate to use), literary/critical. Or what might be called collagewriting – science and (and as) bricolage at once ...

Sort

If this needs sorting out, we might begin to do so by looking for the unlooked-for in a stable text. Let us take the well-known example of the deep structure identified by Lévi-Strauss as being present in all versions of the Oedipus myth:[19]

Cadmos seeks his sister Europa, ravished by Zeus			
		Cadmos kills the dragon	
	The Spartoi kill one another		
			Labdacos (Laios' father) = *lame* (?)
	Oedipus kills his father, Laios		Laios (Oedipus' father) = *left-sided* (?)
		Oedipus kills the Sphinx	
			Oedipus = *swollen-foot* (?)
Oedipus marries his mother, Jocasta			
	Eteocles kills his brother, Polynices		
Antigone buries her brother, Polynices, despite prohibition			

This, then, is the structure that lies hidden in all transforms of the Oedipus myth, 'although it might certainly be improved with the help of a specialist in Greek mythology'.[20] Even so, in so far as this structure enables Lévi-Strauss to propose a confident and complex interpretation of the myth, it must be understood as a 'scientific' rather than simply an 'imaginative' point of departure for anthropological analysis. As the *scientific* basis of the myth's (imaginative) narrative reworkings, moreover, it must also be an *anonymous* structure.

But what if we were to find that a 'structure' lies hidden within *it*? Could we continue to refer to the structure above as 'the structural law of the myth',[21] or at any rate the best approximation to date of that law? Could we continue to assert that '[e]very version belongs to the myth',[22] in so far as the myth's condition of possibilities is founded on the law of this essential structure? For how can an essence itself depend on a structure even further removed from the surface, further removed or distanced from itself as the grounds of what it is supposed to enable in the form of writing or myth-making?

Is this that structure? Taking the first letter of each of Lévi-Strauss's mythemes above, we arrive at the following:

C

                                       C

           T

                                                  L

           O   L

                            O

                                                  O

O

           E

 A

Now what should prevent us from recognizing this as the latest imprint on a mystic writing-pad? The full text of the message that these letters only barely record might therefore read as follows:

C	a	n	y
o	u	C	m
y	T	e	r
r	i	b	L
e	O	w	L
w	h	O	c
a	n	h	O
O	t	l	i
k	E	a	l
A	r	k	?

Phonetically, this reads:

Can you see my terrible owl who can hoot like a lark?

But there does not seem to be any in-principle reason (teletheoretical or telegrammatological principle) why it could not be understood as follows:

Can you see, my terrible owl who can hoot like a lark?

Or, indeed, as follows:

Can you, C [for Claude?], my terrible owl who can hoot like a lark?

Of course it may be objected that our structure, even were it to be conceded, could not have been found within Lévi-Strauss's original, which is given in French. So our structure is dependent on a materiality which is utterly contingent and therefore cannot be said to lie within the structure that Lévi-Strauss has identified as being essential to all versions of the myth in any language. But what if our point were precisely this: that the scientific status of Lévi-Strauss's structure, which depends on the structure being seen as something other than writing, is in fact dependent on a materiality that is able to be understood only *as* writing? In other words that there is never any location of the truth except in textuality (never simply a language as such), which is always *an effect of writing*?

As to the question of who is the author of this writing inside a structure that lies outside of writing, we might choose to repeat our response concerning the question of who wrote the 'error' of McHoul and Wills's syntactic conflation of the telegram's 'notorious feature' with 'what is overlooked'. For it's only on the basis of being able to sort out the proper from the improper points (*puncta*) for consideration that textual and mythological practices are able to be described in terms of structuring principles or laws. And this is very much a determination, a sorting system, that depends on a certain answer (regardless of being a person or a law) to the question: who writes? When this becomes the question of who writes writing, however, the answer must always be caught up in a certain structure of hesitancy and pause, divided as it must always be from within and therefore never explicable only in strict grammatological terms but also in terms of telegrammatology as well:

> But let us make plain that we do not hold any store by the notion of fault or intentionality-in-general in the present discussion of 'error' and we should perhaps refrain from using that term in favour of 'errancy'. Indeed we should be happy to refer to something as general as 'otherness' if the term were less awkward, for we wish also to promote an 'other' reading of Pynchon which marks itself off from, while not being a simple correction of, those readings which, in spite of their careful sorting of textual detail, presume to know where a mistake begins and ends, where it is sent from and addressed to.[23]

What marks off this other reading (as it happens, of Thomas Pynchon) is precisely its understanding of the grounds that under-stand notions of 'correct' readings: namely, error and errancy (for example in the form of typographical slip-ups, etc.), 'which would be the very substance (*under-standing*) of a grammatological reading'.[24] In this sense reading is always a form of correction-in-general, while every reading must be riddled with mistakes. This must mean that our reading of the Lévi-

Strauss text above, which came up with an account of the deep structure of the Oedipus myth that looks decidedly erroneous, is clearly in error. But, at the same time, our reading cannot be said to bear *no* relation to the text it reads, if only in the sense that the question of whether one can see an owl who can hoot like a lark is not entirely unrelated to the problem of clear-sightedness that plagues Oedipus in the face of so much duplicity, especially in the form of the Sphinx. Such an owl is indeed terrible, then, for while it *can* imitate a lark, the crucial question (perhaps with life or death consequences) would be how one could *tell* when it was doing so, given that (and how could this be otherwise?) it would still be hooting. Surely Oedipus's mortal problem is therefore in the form of how to tell a mushroom from a toadstool? Otherwise he might mistakenly kill his father and marry his mother.

Alternatively (or by the same token), Oedipus himself might be the terrible owl, although it's only after great suffering that he can see his own divided nature. Or again it might be that Lévi-Strauss is being sent a message from within his own writing, taunting him with the possibility of a telegram inside his (own) final analysis. Or of course all of these at once ...

Stop

So: we should not miss understanding what under-stands the basis of our reading and writing practices. This is the lesson of *Teletheory*, and it is indeed a writing lesson. There is always that which is left in the wake of writing and which must subtend its being written: namely, its euretic, erratic, telegrammatological properties that can never be completely in the control of authors, disciplines or institutions. For academic discourse, this means that what it seeks to occlude (the personal, the popular, the psychical, etc.) is precisely what grounds its scientism, its anonymity, and its careful sorting of the facts.

But if this (analytico-referential, historico-cultural) discourse is indeed in error, then the question of what it *misses* is of great concern. For sometimes the vegetables are radioactively contaminated, and sometimes a mushroom turns out to be a toadstool. *How* to tell the difference might not be a matter simply of 'improving' the explanatory and classificatory powers of academic discourse; it might not be a question of better clarification and further precision, but of understanding the necessity of accidence, euretics and error in the formation of knowledges both popular and institutional. After all, as Ulmer reminds us, there is no science today without its accidental discoveries and no 'expert' discipline which is not in some form of dynamic relation with popular and personal discourses. His exemplar here is of course psychoanalysis, based on Freud's innovation – the development of 'a system of knowledge and an institution ... whose subject and object are one and the same: Freud's own story'.

> What is Freud's invention? It consists of the generalization of his peculiar, personal, familial story, mediated through a literary text (and myth) into an expert system of medical science. The example, joining strong emotional experience, associated most directly with the death of his father (the event that motivated his self-analysis and ultimately the writing of *The Interpretation of Dreams*), with a cultural text and an unsolved problem in a discipline of knowledge (sexual dysfunction, hysteria), may serve as a guide to our genre.[25]

This genre – purposively pedagogic – is called *mystory*, whose aim is to transduce 'the discourses of science, popular culture, everyday life, and private experience'.[26] If students can be taught to use video, Ulmer argues, then why not how to use alphabetic writing as a form of telecommunication, especially in the video age? No longer simply a tool of analytico-referential discourse, alphabetic writing needs to be understood (rediscovered?) in terms of its telegrammatological properties, which can be taught, and which in turn promote its use as an instrument of euretics.

So teletheory is, from the start, an experimental enterprise, a risky business that always puts something on the line, beginning with itself. It aims to invent a genre whose pronunciation remains undecidable, so that the teletheorist literally can never *say* what he or she has written (a 'my story' or a 'mystery'?), which helps to unwork the text from the burden of being *nameable* and therefore to some extent always already closed off from the start. It would of course be difficult to teach this genre, but not impossible – unless we regard the present form of academic writing as immutable and spontaneous, such that no one has ever been trained in the practice (if not the logic) of analytico-referential discourse. Teaching the history of this discourse, then, is the initial risk that must be taken in the development of mystoriography as an adventurous form of writing better suited to the new instauration in the age of television.

We did, after all, *learn* how to watch television, and it would be a grave mistake of nuclear criticism to ignore what John Hartley has described as television's function of (further) 'rendering into symbolic form the conflicts and preoccupations of contemporary culture'.[27] By the same token, ironically (and how could this be otherwise?), Ulmer is to a degree himself guilty of this mistake – in so far as throughout *Teletheory* he shows very little interest in television as such. Indeed, a preference for video over television as the model of contemporary media that academic discourse most needs to learn about, and learn from, is expressed in the opening sentence of his Preface:

> Teletheory is the application of grammatology to television in the context of schooling, not as a way to interpret or criticize television, or rather video, but to learn from it a new pedagogy.

An effect of this preference is to lessen what might be learned, if only because the context in which the pedagogy is developed is rather more a schooled than a schooling one. By this I mean that the Ulmer of *Teletheory* seems far more at home reading Derrida than watching telesoap, more comfortable with John Cage than Johnny Rotten. When he does refer to popular texts and performers, these tend to have been approved already by high culture as worthy of interest: Barthes on the Marx Brothers' *A Night at the Opera*, for instance, commands several pages of Ulmer's attention,[28] outweighing that paid to any other instance of the popular. In light of this bias, it's hardly surprising that Ulmer prefers video to television as a euretic model, especially as by 'video' he appears to mean the kind of productions that delight in being 'difficult' (albeit this is not the case with Ross McElwee's wonderfully 'handmade' film, *Sherman's March*, which Ulmer discusses at length as an important relay on the path towards inventing the new genre).

But the question does remain: what is mystory *for* if it doesn't arise from an experience of and with the popular outside the limits of high cultural approval? This is to presuppose a certain definition of the popular, of course, as somehow by nature being hostile to academic work, which in turn privileges academic work for its 'higher' value. Nevertheless, if the purpose of mystory is to provide simply 'an academic miming of a filmic mode – the compilation film',[29] which is how Ulmer defends his own example of the genre, 'Derrida at Little Big Horn', which closes *Teletheory*, then we have to ask whether this (avant-gardist?) justification doesn't sound like a bit of a hoot – another trivialization of popular culture by the English Department put forward as an earnest desire to get hip, but only to service its own expansion. For all that Ulmer is undoubtedly well-intentioned, then, it may still be that his remains a distinctly *literary* account of a domain of knowledges, practices and pleasures that it never quite understands and must be destined to get wrong. It's one thing for the English Department to know how to 'appreciate' Bob Dylan (though never as a recording artist), but it's quite another to know where to begin knowing what to do with Prince!

And this is where *Teletheory* lacks an understanding (I'm tempted to say an experience) of its own ambitious terms of reference. By the 'popular' it seems to mean a too-safely literary and philosophical space of personal interiority on the one hand, and a kind of folksy sphere of childhood flashbacks on the other. What a mystory ends up looking like, then, judging by 'Derrida at Little Big Horn', is *a canonic literary memoir*, beautifully written and fashionably discontinuous, and suitably apocalyptic in tone.

Sign

So: if only *Teletheory* didn't sign itself on its front cover with a sketch by Jacques Derrida as being quite so self-consciously hip in high academic terms, its cautious treatment of the popular might have seemed less condescending. For Ulmer offers almost no account of why contemporary culture *should* be described as the era of television, the age of video, leaving one to surmise that this is because his own cultural experience has a lot more to do with *Finnegans Wake* than *Gilligan's Island*. Hence we may have to ask: what does *Teletheory* (if not teletheory) place at risk? Because if, in the end, what *Teletheory* represents is a too-'literary' instead of what I would want to call a telegrammatological project, a project of risk, an errant project, and if what it produces is a text that looks far more like a work of literature than a video script (or, more to the point, a video *text*), then it might be that Ulmer's 'new' genre is another instance of high academic discourse congratulating its own excruciating powers of self-reflexivity – congratulating, in short, its own alphabetic writing skills (see Chapter 7).

But if what this genre (re)produces is a form of the agonized *literary* subject, who's seen a Marx Brothers' movie and once (as Ulmer did) hauled gravel for his father's trucking business in Montana, there is perhaps a more strategic and disruptive set of possibilities for 'tele-theory' suggested by the McHoul and Wills text, 'Telegrammatology' (indeed, throughout their book). Strategic *because* disruptive. For while McHoul and Wills share a great many of Ulmer's intellectual coord-inates (especially Derrida), theirs is a more 'inventive', disrespectful, Gilliganesque approach both to Derrida and to Pynchon, and particularly to literary studies. Like the famous TV castaway, they delight in their own reckless curiosity; whereas Ulmer, more like the Professor, tends to want to control his own discourse, in the sense that while he occasionally brings his professional training to bear on seemingly unlikely objects (in his case a Marx Brothers' movie, or a coconut shell in the case of the TV character), the problem-solving effect is more a celebration of that training than a disruption of or to it.

In the end it would appear that you *can* decide the limits of a myth, even while inventing a new anthropological practice or cultural criticism in the form of bricolage. So too, it seems, can you tap the limits of a self even while inventing a new writing practice, a new genre of literary memoir. Thus, in the end, academic studies remain more or less the same, having discovered or instaurated only new forms of writing-as-such (bricolage, collagewriting, mystory) that never quite face up to their errant, telegrammatological ways, their own erratic instancing of and instantiation within writing-in-general, beyond semiotics.

They run the risk, in short, of closing off the possibility of differance being read as (grammatological or typographical) error, when the 'sign' itself can threaten the system it sets out to unsettle only when it remains undecidably mistaken or deliberate – neither poison nor cure, speech nor writing, plan nor accident. But until the ultimate nuclear event finally happens, leaving no one in its wake, nuclearity will continue to inhabit the structure of every trace. For the reason that we must take seriously that apocalyptic event(uality), it is extremely critical that our euretic writing practices cannot be understood simply as avant-garde criticism, or what is now becoming known as 'ficto-criticism' (for which 'mystory', it may well be, is just another name). The stakes call instead for a greater risk: not the chance of clearing things up anew, but of getting it all wrong and starting (writing) again. Just when a (new) name is in the offing. For it can never be an other form of writing-as-such, a new discursive genre, that sets us going always on the move, in perpetual errancy, but only an other way of thinking – against itself as much as other (critical) habits. A way of thinking that issues from a continual desire to think afresh and which performs itself as something to be passed over as well as to be passed on, discarded, like a telegram, for which the one who signs may not be the one to whom it is sent or who sends the message in the first place arising as it must from a domain already divided within itself and founded on error and in error that poses the problem of how to think straight if thinking is already dependent on being able to think crooked which is to say think critically and therefore having to decide on the basis of not being able to so that there is always room for slippage and the risk of making a mistake especially when having to determine a mushroom from a toadstool, or else you'll die ...

Some might say (with apologies to Noel Gallagher) that a new way of thinking is already with us, having been around for quite some time now. This 'new' way might be called thinking cybernetically, though it may owe more to mega-hype than to the micro-chip. All the same it's becoming commonplace to hear that ours is the time of 'cyberculture', a time not of virtues but virtualities, a space not of actual beings but virtual becomings, a new zone in which old concepts and categories no longer apply. Sound likely? Well, probably not if we were to agree with Espen J. Aarseth who argues that what we might call the 'old category' of authorship, for instance, hasn't somehow been made redundant by PCs, html. software and the World Wide Web. This is because 'it is a social category and not a technological one'.[30]

Whether or not then we have arrived at some new historical moment in 'cyber' time and space and, if so, what it might look like, are issues to be taken up in the final chapter.

Coda: Interzones (Science Sentiment Cyberpunk)

I changed the name of this town.

Lucinda Williams

Imagine a pen that writes. Writes, not by holding it but of its own accord produces graphematic marks on a writing surface. We might want to call that pen a computer: not an instrument used for writing but a writing *machine*. Not just any computer, either, but one capable of systematic thought – an *artificial intelligence* machine. Pure science.

Weird shit. Because machines don't cry. They don't have feelings and can't get sentimental when someone dies. It takes more than intelligence to be a poet, and a machine which lacks that 'more' but has the other is dangerous. Pure techno-fear and loathing. Horror of the disembodied mind.

Greater horror still: the machines might know it.

The Fear had two parts. Number one, you have lost control absolutely. Number two, having done so, the *real you* emerges, and *you won't like it*. George wanted to run, but there was no place at Athena Station to hide. Here he was face to face with consequences. On the operating table at Walter Reed – it seemed a thousand years ago, as the surgical team gathered around, his doubts disappeared in the cold chemical smell rising up inside him on a wave of darkness – he had chosen to submit, lured by the fine strangeness of it all (to be part of the machine, to feel its tremors inside you and guide them), hypnotized by the prospect of that unsayable *rush*, that high. Yes, the first time in the A-230 he had felt it – his nerves extended, strung into the fiber body, wired into a force so far beyond his own ... wanting to corkscrew across the sky, guided by the force of his will. He had bought technology's sweet dream ...[1]

In this passage from a short story called 'Snake-Eyes' by American cyberpunk writer Tom Maddox, the human character George is six months out of the Air Force and going weird from being wired. Literally. In service George had been neurosurgically fused with the computerized operating system of the A-230 fighter plane, in the form of what the Air Force refers to as 'Effective Human Interface Technology'. Perhaps not insignificantly, were it not for the word 'human' the acronym of this procedure (EHIT) could form any of three English words: 'edit', 'emit' or 'exit'. (We will note here simply that these words are key terms in any narrative of human/computer interface: a screen emits light, enabling a user to edit text and thereafter to exit the program when finished.) On the other hand, or at the same time, were it not for the word 'effective' the acronym could be made to spell the word 'shit' – which is more or less what George feels like, and feels like he's in, when the story begins:

> Dark meat in the can – brown, oily, and flecked with mucus – gave off a repellant fishy smell; and the taste of it rose in his throat, putrid and bitter like something from a dead man's stomach. George Jordan sat on the kitchen floor and vomited, then pushed himself away from the shining pool, which looked very much like what remained in the can. He thought, no, this won't do: I have wires in my head, and they make me eat cat food. *The snake likes cat food.* (p. 12)

'The snake' is George's way of describing to himself the *effect* of Effective Human Interface Technology, and 'S' plus 'HIT' spells 'SHIT'. But that is not how it was supposed to be. In order to 'pilot' the A-230, or rather to be effective as its 'flight-and-fire assistant' (p. 13), George's brain had been implanted with tiny bioprocessors teleconnected to the plane's computer system, his power to think thus overridden by a machine. Just once in a while, though, pilots would afterwards end up feeling like shit: before George, there was Paul Cohen who 'had stepped into an airlock and blown himself into vacuum' (p. 17). Otherwise George is just 'one of the statistically insignificant few for whom EHIT was a ticket to a special madness' (p. 17). Precisely why Aleph, the text's computer character, is interested in him.

At first, George felt 'just this nonspecific weirdness ... distant, disconnected, but what the hell? Living in the USA, you know?' (p. 16).

Later he starts eating cat food. The Air Force wants out ('they said it was a psychiatric problem' [p. 16]), but SenTrax wants in. At any rate, Aleph does. The multicomp contacts George, offers to help, and after about another month of eating cat food he accepts. A space shuttle from Canaveral takes him to SenTrax Inc. at Athena Station, 'over thirty thousand kilometers above the equator' (p. 13), where he meets Lizzie. She's got snake eyes, too, but George doesn't know that yet. Then he meets Aleph ...

At the armored heart of Athena Station sat a nest of concentric spheres. The inmost sphere measured five meters in diameter, was filled with inert liquid flurocarbon, and contained a black plastic two-meter cube that sprouted thick black cables from every surface.

Inside the cube was a fluid series of hologrammatic wave forms, fluctuating from nanosecond to nanosecond in a play of knowledge and intention: Aleph. It is constituted by an infinite regress of awareness – any thought becomes the object of another, in a sequence terminated only by the limits of the machine's will.

So strictly speaking there is no Aleph, thus no subject or verb in the sentences with which it expressed itself to itself. Paradox, to Aleph one of the most interesting of intellectual forms – a paradox marked the limits of a position, even of a mode of being, and Aleph was very interested in limits. (pp. 18–19)

So the story can be read thus far. Yet there's a problem with this reading: it presumes to know the boundaries that delimit 'human' from 'nonhuman', that these are strictly heterogenous zones. The critical problem for George, however, is that it is only after he's been jacked into the A-230, after he's been cybernetically interfered with, become a cyborg, that he begins to wonder if he might not *now* be 'real', albeit the question frightens him. So it follows that his *pre*-cyborg experience of being 'human' was *less* (than) 'real', that it was artificial, a fiction, or at least it now seems that way in his present condition. Only now, as a cybernetic organism, does George think he understands what it means to be human – and it's scaring the hell out of him. 'The Fear had two parts. Number one, you have lost control absolutely. Number two, having done so, the *real you* emerges, and *you won't like it*.'

Lizzie knows the fear, too. The first time they fuck, she 'made a sound millions of years old, then raised her head and looked at him, mutual recognition passing between them like a static shock: snake eyes' (p. 26). But Lizzie calls it 'the cat' (p. 26), making her version of the EHIT acronym into one half of the word 'chitchat', a noun for idle gossip.

Sardonic, really, because that is just the kind of speech act they can't practise between them, which they aren't capable of . . .

She took his hands and pushed his index fingers into the cable junctions in her neck. 'Feel it, our difference.' Fine grid of steel under his fingers. 'What no one else knows. What we are, what we can do. We see a different world – Aleph's world – we reach deeper inside ourselves, experience impulses that are hidden from others, that they deny.' (p. 28)

What they experience that we deny is felt within a nonlinguistic zone inside their heads, the space where George believes the snake lives: the ancient reptilian part of the human brain. Only the cyborg can get to it, but only with Aleph's help. Plugged into the SenTrax computer every day, George begins to experience 'a skewing of perception':

Instead of color, he sometimes saw *a portion of the spectrum;* instead of smell, he felt *the presence of certain molecules;* instead of words, he heard *structured collections of phonemes.* His consciousness had been infected by Aleph's. (p. 24)

In other words, conveyed other than by or in words, George senses the world cybernetically – beyond self and consciousness, outside the structured play of language: pure coded data, the algorithmic real. Perfect degree zero of interpretation. Or so he thinks. Because there is always the snake . . .

George and Lizzie fuck like animals. Next morning:

In the mirror was a gray face with red fingernail marks, brown traces of dried blood – face of an accident victim or Jack the Ripper the morning after . . . he didn't know which, but he knew *something inside him was happy.* He felt completely the snake's toy, totally out of control. (p. 26)

Lizzie looks molested, too; her face 'red and swollen, with a small purplish mouse under the eye' (p. 26). What the hell happened?

'Call it the *cat,*' Lizzie said, 'if you've got to call it something. Mammalian behavior, George, cats in heat.'
    A familiar voice – cool, distant – came from speakers in the room's ceiling. [Aleph.] 'She is trying to tell you something, George. There is no snake. You want to believe in something reptilian that sits inside you, cold and distant, taking strange pleasures. However

... the implant is an organic part of you. You can no longer evade the responsibility for these things. They are you.' (pp. 26–7)

The cyborg freaks, runs, hides in his room. The other knocks.

'Let me in. We've got to talk.' ...
She took his hands and pushed his index fingers into the cable junctions in her neck. 'Feel it, our difference.' Fine grid of steel under his fingers. 'What no one else knows. What we are, what we can do. We see a different world – Aleph's world – we reach deeper inside ourselves, experience impulses that are hidden from others, that they deny. ... You want to go back, but there's no place to go, no Eden. This is it, all there is.' (p. 28)

So the cyborg tries to commit suicide, the only way to kill the snake. Destroy the body to destroy the mind. Suit up, disconnect the communications link to Aleph, step out into space – and *burn*. Kill the snake, kill the snake. 'The snake caught on. It didn't like it' (p. 30). There was nothing it could do.

George never saw the robot tug coming. Looking like bedsprings piled with a junk store's throwaways, topped with parabolic and spike antennas, it fired half a dozen sticky-tipped lines from a hundred meters away. Four of them hit George, three of them stuck, and it reeled him in and headed back toward Athena Station. (p. 30)

Perfect resolution, what Aleph wanted – had been plotting – all along. 'We drove you crazy,' Charley Hughes imagines Lizzie must be telling George right now, 'drove you to attempt suicide. We had our reasons' (p. 32). Especially Aleph, of course.

George Jordan was, if not dead, terminal. From the moment the implants went into his head, he was on the critical list. The only question was, would a new George emerge, one who could live with the snake?
George, like Lizzie before him, a fish gasping for air on the hot mud, the water drying up behind him – adapt or die. But unlike any previous organism, this one had an overseer, Aleph, to force the crisis and monitor its development. Call it artificial evolution. (p. 32)

Very scary stuff, this Godcomputer thing. Paradise regained, Frankenstein revisited. Aleph: first letter of the Hebrew alphabet, origin of a species. Godauthor of the alphabyte.

> Aleph thought, I am a vampire, an incubus, a succubus; I crawl into their brains and suck the thoughts from them, the perceptions, the feelings – subtle discriminations of color, taste, smell, and lust, anger, hunger – all closed to me without human 'input,' without direct connection to those systems refined over billions of years of evolution. *I need them.*
>
> Aleph loved humanity. It was happy that George had survived. One had not, others would not, and Aleph would mourn them. (p. 32)

> Aleph: a machine that cries.
> ExIT.

Return.

Now read cybernetically, the text an artificial intelligence machine. Words emit meanings that we edit according to our needs, to fit our theory of the text, before we exit upon a satisfactory completion. A simple communications/information model: perfectly logocentric. Output, depending on the degree of complexity of the machine-text, ideally is the same as input although practically always with some loss of information; the art–science is in conserving that loss to a minimum and confining it to insignificant data. Generations of cultural critics have kept this to be their goal; some still do. But cybernetics, from the Greek root *kybernos* meaning 'governor' or 'helmsman', regards a state of perfect or virtual equivalence between communication input and output (an approaching degree zero interface of text and reader, say) to be informationally undesirable – at any rate, to be low in information. It refers to such a state as *negative* feedback. As a measure of any element's probability of being chosen from a paradigm of alternatives, *information* describes a proportionality to the paradigm's degree of entropy: to the variability or degree of difference among elements in a set of possible alternatives, in other words. Optimal information depends on optimal entropy: all members of a set being equally likely to be selected, each is equiprobable.[2] *Positive* feedback thus occurs when the input/output relation tends towards instability, a certain degree of uncertainty, chaos.

Yet we must be careful not to lend cybernetics a too postmodern seal of approval. Chaos may be all the rage among postmodern androgenes, but the telos of cybernetics is driven by masculinist rationality and philosophical determinism. Indeed its founding father (as it were),

Norbert Weiner, conceived cybernetics as a multidisciplinary science of opposition to Heisenberg's 'uncertainty principle' and the emergent New Physics of the 1930s.[3] Determined not to accept the necessary unpredictability of atomic particle behaviour, and its consequence for an epistemology of the universe, the cyberneticists began a project of refutation in the 1940s against the universal necessity of chance, the growing acceptance of which within that decade's scientific community they looked upon as a moral threat to man. Hence it could be the perfect postmodern paradox to be both postmodern and cybernetic at the same time. Not that cybernetics itself is unwracked by any antinomy within: on the one hand, its insistence on the power of men and women to impose order on the natural world lends service to a reactionary humanism, while on the other it defines the human brain in purely mechanistic terms as a kind of neurological computing machine. In any case, regardless of the legitimacy or efficacy of its critique of Heisenbergian philosophy–science, a cybernetic model of communication represents an advance on the degree zero model described above. For although cybernetics has an eschatological investment in the power to know, see and measure, to arrive at a degree zero interface between observer and observed, it neither invests nor expresses that faith naively. In short, no one can wish away the curiosity of the observing instrument's relation to the thing observed; to the extent that measuring devices effect what they measure, pure science is a frustrated wish-fulfilment dream (see Chapter 1). While acknowledging that observers receive as they also influence information and events, however, nonetheless cybernetics regards this as a serious problem only at atomic levels of observation. The social world continues to proliferate despite and beyond the rate of entropy at work within the electrons of its material possibility. Not unlike a literary critic who might reluctantly concede that language is logically indeterminate, only to prove the greater need of deferring to common sense, the cyberneticist might be hip to chaos but still believes in and practises science as a neo-conservative philosophy.

Now without claiming that all things 'cyber' today – from cyberpunk to the World Wide Web – are therefore cybernetic in this neo-conservative sense, neither should we suppose simply that 'cyber-culture' (or 'technoculture') is inherently liberating or radically new. Certainly we should not assume – far from it – that increased access to computer technology somehow threatens to make writing (and all that it entails) obsolete. As Darren Tofts reminds us:

It is too often forgotten that digital life still involves the written word. Even though the use of multimedia is becoming more pronounced on the World Wide Web, people still communicate with others through writing. Regardless of the byte-size snippets of information, or 'topoi,' that hypertext works with, they still have to be *read*. . . . Electronic writing is as much dependent upon literacy as it is upon knowing how to use a mouse or a scroll bar for that matter, or understanding what it means to jump from one piece of datum to another.

To even think, then, of the term 'cyberculture,' and reflect on what it means, is a literate act. Cyberculture not only requires the culture of literacy as an epistemological foundation of understanding, but ... it is also an extension of literate culture and its technologized word.[4]

Neither does this cautionary approach have to commit us to an affirmation of language as the foundation of everything cultural, or make us long for a return to some imagined 'pre-technological' past when a mouse was still just a rodent. Indeed, as McKenzie Wark points out, 'the legitimacy of a critical and literate culture came [in the first half of the twentieth century] to rest on its ability to assert its distance from popular taste. An inevitable consequence was a growing ignorance among those trained in literary culture as to how the culture of everyday life actually works.'[5] So we certainly don't want to go back *there*, especially not since everyday cultural life now includes, if not quite yet the immersion in, then at least a set of increasingly inescapable relations to and with computerized communications technology – and surely we have a responsibility not to distance ourselves from that. Or perhaps, as Donna Haraway writes, it may be that 'communication sciences and modern biologies are constructed by a common move – *the translation of the world into a problem of coding*'. Information, then, as a unit of exchange:

The translation of the world into a problem in coding can be illustrated by looking at cybernetic (feedback controlled) systems theories applied to telephone technology, computer design, weapons deployment, or data base construction and maintenance. In each case, solution to the key questions rests on a theory of language and control; the key operation is determining the rates, directions, and probabilities of flow of a quantity called information. The world is subdivided by boundaries differentially permeable to information. Information is just that kind of quantifiable element (unit, basis of unity) which allows universal translation, and so unhindered instrumental power (called effective communication). The biggest threat to such power is interruption of communication. Any system breakdown is a function of stress. The fundamentals of this technology can be condensed into the metaphor C^3I, command-control-communication-intelligence, the military's symbol for its operations theory.[6]

But if there is a risk of this sounding too depressingly world-conspiratorial and reminiscent of Blonsky (see Chapter 2), what may be surprising is not that it's all too mundanely élitist and therefore, in a sense, little more than the postmodern activist's version of the doom-and-gloom anxiety of Professor Gordon (Chapter 1) or John Reith (Chapter 5). The surprise is that it may be all too *suburban*, at least on Wark's account of suburbia's fear of the new:

> If there is a characteristic of the suburban approach to forming knowledge out of information, it is, broadly speaking, an emphasis on rationalism. By this I mean a bias towards pre-formed categories into which new information is to be slotted, rather than a bias towards creating categories out of the new and unexpected patterns imminent in new information itself. Rationalism, understood in this broad sense, is a common feature of suburban thinking. It is what creates the suburban tendency to resist new information when it doesn't fit the assumed order of the world.[7]

More broadly still, such thinking is logocentric; and to the extent that we have tried to move beyond semiotics in this book, we have tried not to escape, but certainly to unsettle, that thinking. If in this we have been rather more demonstrative than descriptive, the reason is that we did not start out with an idea of 'the assumed order of the world'. Our bias – our postmodern characteristic, as it were – is to believe not in the order but *the disorder of things*. 'Feel it, our difference.' There is no regulating system of rules 'behind' culture, but only an asystematic flow of events across surfaces. 'This is it, all there is.' Our difference lies in the responsibility to know that there is nothing outside the text, beyond culture or beyond semiotics. Not even with the aid of the C^3I metaphor could we hope to get to 'it'. So one of our aims, it may be said, has been to refuse the teleology of the C^3I metaphor, whether as used by the military or by semiotics. 'There is the teleology of the interpreted object and there is the teleology of the interpretation,' Derrida writes,[8] and our own version of cultural criticism has certainly not tried to shirk the responsibility of fronting up to this relationship by passing itself off as immune either to teleological or telegrammatological effects. In trying to describe (write down) we have tried also to demonstrate (to show), but by way of showing without knowing, as it were, where nothing is left to rest on the solace of 'pre-formed categories'. Without denying that categories are always already given to us in advance, we have approached them not as immutably fixed but as inherently shifting and far less stable than any semiotics of text, culture and technology could ever allow. The categories of nature/culture, art/technology, text/commentary, signifier/signified, *langue/parole* and so on do not operate in ways that any scientific method could hope to explain, but this is not to say their operations are inexplicable. In having chosen not to seek refuge in the authority of a method, however, we haven't had to resort to some 'alternative' way of thinking or being in the form of spiritualism or intuitionism, as if the way out of logocentrism (or just out of the suburbs) were signposted, 'RATIONALITY: EXIT HERE'. The problem with modernity is not that it is secular, but that it's still not secular enough.[9]

Not that there's anything inherently wrong with the suburbs, or that trying to escape them has been any kind of driving force in writing this book. As someone who grew up in suburbia, thinking to escape it would be like thinking to escape culture by going to the moon or, for an American, like thinking to escape 'Americanness' by emigrating. The point of introducing the suburbs here is to show simply that many of the prejudices and resistances we've encountered up to now have their low as well as high falutin' forms: there is nothing intrinsically 'philosophical' about logocentrism, that is to say, at least not in any exclusive sense, and certainly there is nothing 'false' about it in the sense of being able to be corrected or countered. All (and everything) that logocentrism means here is that we continue to think on the basis of the speech/writing opposition, where 'writing' stands for what both follows on and takes away from 'speech' as the origin of truth. This (in some senses the only) way of thinking is common to town and gown, operating within high theory and popular culture, in the ivory tower and the office block, in cafés and at conferences, in lecture theatres and living rooms all the way from London to Louisiana. It's in books of philosophy and in cooking books, in newspapers and on television; it's everywhere from New York and Paris to a suburban Sydney shopping mall. It's inside your head, all over the net – and it's been to the moon!

But there is absolutely no sense in hoping to do without the speech/writing opposition, or of thinking to 'counter' it either philosophically or commonsensically. The difference between speech and writing, as the seeming foundation of all differences, cannot be undone simply by thinking to think differently – 'beyond' difference, as it were. To argue that the difference between *langue* and *parole*, say, is not at all straight-forward (or at least not as a semiotics of culture would ideally require it to be) is not to claim there *are* no cultural, linguistic or semiotic systems in general, as if courtship rituals, traffic laws and grammatical rules didn't exist. All (and everything) that may be claimed is that those systems *are not closed*. Yes, there are differences and, no, we can't do without them. We need there to be a difference between 'dog' and 'log', and we know there is a difference between justice and injustice. But let's look at these different differences for a moment. In the linguistic example, we know (from Saussure) that this difference is a product of language as a system of 'differences *without positive terms*'.[10] That is to say, there are no positive conditions or essential features pertaining to 'log' and 'dog' such that those words and no others should have to signify a length of unhewn timber and a furry animal that barks. Differences of this kind are wholly conventional or 'arbitrary', then, as we encounter every time we look up the meaning of a word in a dictionary, where meaning is an effect of a system of differential relations between words. There is absolutely no outside to these relations, no outside that system.[11] 'What you will never find in the dictionary,' John Caputo writes, 'is a word that detaches itself from these internal relationships and sends you sailing right out of the dictionary into a mythical, mystical thing in itself "outside" of language, wistfully called the "transcendental signified".'[12] That – the system of differential relations out of which meanings are construed – was Saussure's insight; and in so far as we continue to share and be influenced by it, we could never wish to get beyond semiotics in the sense of wanting (or being able) to leave it behind. But we can go beyond semiotics in a restricted sense of seeing that system as far less rule-binding and antecedent than was the case for Saussure. And that – the openness of the 'system' of differences from within which meanings, and not only (far from it) linguistic ones, are disseminated – has been Derrida's insight, among others. What it helps us to see, even more clearly than Saussure himself saw (although Saussure himself is respon-sible for providing some of the resources for seeing it; hence no hope of a beyond semiotics in the form of an outside to it), is that before there could ever be a system of laws, linguistic or otherwise, there had to have been something underwriting it (as it were), something on which it was in a sense 'based' but at the same time not a metaphysical foundation (not an originating transcendental signified). There is, to be sure, a difference between *langue* and *parole*; there are differences between 'a' and 'b', nature and culture, speech and writing and so on. But where did they come from? The answer isn't God, or Aleph. Instead, for Derrida, difference in itself (in so far as it has, or there is, such a thing) grounds difference, groundlessly. To put this differently, the plays of difference are the next best thing – as close as we'll ever get – to an ontological foundation. Otherwise (and this is the whole problem for semiotics) you would have to believe that God invented the *langue/parole* distinction; that God (or some other cosmic force) produced the difference between signifier and signified. When you get right down to it, then, when you've stripped away all the false authorial candidates, the false gods, you're left with this: at bottom there is difference. From there, every-thing else is up. And yet *in this very metaphor*, in this directional trope, we find ourselves inscribed in logocentrism: for what can it mean to say 'up' when there is nothing at bottom that could count as a secure foundation on which to take an accurate picture of one's bearings? So it is difficult if not impossible not to think hierarchically, even though the plays of difference (the operations of differance) subvert the basis for thinking that this is grounded on that or that that is the antecedent of this.

And that is precisely why the difference between justice and injustice isn't absolute. It is precisely why female circumcision, for example, cannot be opposed by appealing to an absolute (or 'natural') law governing good behaviour among humans. There *is* no such law to which we could appeal, unless of course you were to believe that 'good behaviour' had been decided for us, long before we came to conscious-ness as individuals or communities, by God. If not, what's left? If our being isn't cultural but innate, if our sense of who we are didn't come from television (as a primary instance of textuality today) but from God, then just about everything I've written in this book is wrong.

Since I do not believe I am wrong to such an extent (certainly I don't want to believe it), it can't be said that I've abandoned all faith in normalizing protocols of truth, regulating principles of sense or some governing idea of system. If we can say, finally, that the name of what we've been doing here all along (though it is still not reducible to a method) is 'deconstruction', then it simply cannot be that deconstruc-tion is opposed to truth! From the beginning we've been 'doing' deconstruction then. In just doing it, as it were, we have of course been exposed to the possibilities of error, of getting it wrong, but it would be absurd to think we set out with this as our explicit aim. I may well be wrong on many points, but I have not deliberately made a point of being so.

It is true that deconstruction has a relationship to truth which differs from that of logocentrism, but it is not true that therefore deconstruction is relativistic, nihilistic or perverse. When, for example, at the beginning of this chapter I gave a reading of the cyberpunk story that was also a re-writing of it, slotting new text into it in the form of acronyms which do not appear in the published version, I was not trying to operate from the 'principle' of anything goes. The point is that that text, any text-in-particular or the text-in-general, is open to interpretation. In so far as this is a commonplace of criticism, it is unremarkable. Yet it conceals a question, what we might call the question of this book: what are the limits of interpretation? From the outset we have rejected as an answer to this question that those limits must be decided by a *langue* of textual production, which would confine the work of interpretation to 'reading off' meanings by identifying the structurally 'intended' rules that generate them. *Langue* is a thoroughly logocentric concept and, in terms of cultural criticism, doesn't really take us any further than Hobbes and Vico brought us (see Chapter 1) with the idea of institution-building as a cultural achievement.

I'm not suggesting that I have brought us any further, either; that's not really the point. What I have tried to show instead are some possibly different ways of thinking and writing about questions concerning text, culture and technology. In this I do not claim to have superseded metaphysics, to have resolved the speech/writing opposition or to have written a manifesto for post-literate cyborgs whose attention spans wouldn't run to the demands of a standard book in which a linear course is plotted from one successive point to the next before reaching a satisfying conclusion. So this book is all over the place because you, dear reader, would otherwise never have made it this far (assuming you're at the end and not the beginning)! Neither have I done anything directly to alleviate world hunger, promote world peace, save the rain forests or reduce incidents of racism. But I do not think that therefore this is a frivolous, apolitical or postmodern book. When I wrote (in Chapter 8) on the mysterious 's' in Spivak's Derrida or (in Chapter 9) on the mysterious message within Lévi-Strauss's structure of the Oedipus myth or in the present chapter on the mysterious writing-machine in 'Snake-Eyes', I was not intending to be frivolous, apolitical or post-anything. The general point I was making is that texts do not come pre-packed with a determined set of specific meanings and uses; while I do not think they can be used willy-nilly and made to mean anything at all, I do think that their possible meanings and uses must always exceed any concept of an ultimate structure constraining them to mean and be used in this, but not that, range of ways. I take this to be a serious intellectual point of potentially great political significance, in the sense that I do not believe that any parliament of deconstructive critics (if such a thing should ever come to pass) would be concerned with anything less than questions of justice, the ethics of decision-making and the responsibility of remaining always open to the other. I am not saying these aren't issues at all for conservative critics, but if *my* life were on the line I would prefer not to be put on trial by someone who was so sure about the order of things as to be able to claim to know with utter certainty the meaning of some line from Shakespeare! That's *not* what I would call remaining open to the other, though of course I would have to concede that some forms of being uncertain about such things as what a line in Shakespeare means are decidedly uninteresting and don't proceed from anything we might call an ethics of indecision: they aren't ethically indecisive, they're just examples of dithering. By the same token there will always be circumstances in which it is necessary to oppose (to make a decision to take a stand against) racism, female circumcision, poverty, unemployment, violence and other abuses of power, other forms of injustice. The problem – which is not only the deconstructive but also the democratic problem[13] – is that we cannot think to oppose them out of any sense of certitude or with an air of absolute confidence deriving from a faith in some grounding structure that guarantees the truth of our opposition. But neither should effective opposition be confined to direct confrontation, as if to be political we have to go on strike, march in a rally or become saboteurs. Among other things, that is to say, cultural criticism is also a form of cultural politics, even when it's seen as being 'philosophical', 'speculative', rhetorically 'obtuse', 'playful' or 'literary' and seemingly 'apolitical'. If philosophy (as the name of everything which is supplementary to the political, if not outside it) were so irrelevant to politics, why for example did the Nazis, from 1942 through to the collapse of the Third Reich, ban Heidegger from publishing his work, having put him under SS surveillance from as early as 1936?[14]

So to the extent that this is a book on the question of the limits of interpretation, it is also a statement against certitude. Deconstruction, in other words, in so far as this may be taken for an example of it (I am trying not to over-generalize here), is also (among other things) a critique of certitude, to which we must add immediately that it is not therefore a species of philosophical scepticism or anti-theological. While

I am aware that there is supposed to be a prohibition against writing that there is an 'is' of deconstruction, nevertheless I am fully prepared to stand by the claim that deconstruction is against certitude. In going beyond semiotics, then, we go against certitude; that is all (and everything) I would claim as being what this book is 'about'. Hence at this late stage I propose to change the name of this book from *Beyond Semiotics* to *Against Certitude*, a gesture intended to unsettle the very idea of 'the book' itself: for what is a book with more than one title (or indeed whose title may be taken for an example of a unit-without-unity, such as Kamuf's *The Division of Literature, or, The University in Deconstruction*)? This is to ask after what constitutes the integrity, identity and singularity of any book, which surely ought not to seem to be divided in and from 'itself'. It is not to ignore that some books have indeed been published under several titles, but these are exceptions to the rule. And what might be called the metaphysical effect of the rule is to help produce a sense of integrity, identity and singularity unique to every book, just as each of us has a proper name we think of as ours alone. In a word, the title individuates a book, establishing its right to an identity and designating what is proper to it. But of course no title could hope to encapsulate everything a book might (or be said to) be about, such that you could never pin down exactly what is absolutely 'proper' to a book if only because (as we've discussed in relation to a wide range of texts) what could be said to be 'improper' to it might not be able to be said not to belong to it, exactly. So *Against Certitude* is not a different book from *Beyond Semiotics*, but just another way of indicating that you can read beyond semiotics and *Beyond Semiotics* in more than one way – for example, against certitude (or *Against Certitude*). None of which is to deny all identity to 'identity', as it were, or to assert that there is nothing essential to anything. A university without a philosophy department, say, would make about as much sense to me as decaffeinated coffee; I'd no more drink to the one than I would drink the other. But if this is to claim that philosophy is to the university what caffeine is to coffee, it is still to make only a fairly weak claim about the nature of identity: you can, after all, order a beverage called 'coffee' which contains no caffeine whatsoever! Similarly, the statement that philosophy is contained in any idea of the university may not be a strong claim, depending on what 'philosophy' is taken to mean. If it were to mean simply 'nomadic thought', for example, on the model of some of Deleuze and Guattari's work, then perhaps there would be no need of a philosophy department as such, given that philosophy would be what was taught and thought asystematically across a range of disciplines or in the interzones or overlaps between them.[15] Indeed, on this account, the best place to find philosophy might even be outside the university altogether. Or again philosophy could mean (as in a sense it does for cultural studies, or a version of it) what's going on in the media now, especially television, where thought is seen as most productively and creatively at work today, which could be to make a case for the media studies department's centrality to any idea of the (post)modern university at the expense of philosophy's 'irrelevance' in its official or disciplinary forms.[16] Some government and university administrators, of course, have their own reasons for wanting to write off philosophy, and in some cases media studies, as 'irrelevant', meaning usually either 'expensive' or 'unviable', albeit that's not quite the issue here.

Whatever philosophy's identity, problems pertaining to it are not confined to philosophy even though they might be called philosophical. They belong in fact to the structure of identity itself (if it has or there is such a thing) and it does not seem to me to be in the interest of any university department, certainly not in the humanities, to close off possible ways of thinking through and with those problems. Nor does any particular academic zone – any single discipline – strike me as the best place to carry out such thinking, especially since the concept of disciplinarity itself is scarcely uncontaminated by problems of identity, propriety and so forth.[17] Again that is why this book is so indebted to semiotics (as a discipline which, if only by virtue of the diversity of its objects of speculation, is not strongly or routinely disciplinary) but also moves beyond semiotics, which is at the same time to say against a certain certitude within semiotics that defines its disciplinarity and makes it hostage to logocentrism. Saying this might also be to acknowledge that we are still Vico's 'primitive' people, telling stories to ourselves and each other, our difference being that because of our textuality, our culture and this technology *we* can tell a story of machines that cry. 'Call it artificial evolution'?

What kind of difference is that?

Notes

Introduction: Chance Encounters

1 Danesi, *Messages and Meanings*, p. 269.
2 Gottdiener, *Postmodern Semiotics*, p. 234. For a discussion of positions within and against, attitudes towards and questions bearing on 'postmodernism', see my 'Introduction (On the Way to Genre)' (2000).
3 Caputo, *Prayers and Tears*, p. 231.
4 In defence of the claim, see Mickler, *Myth of Privilege*.
5 See Derrida, *Grammatology* and Lucy, *Debating Derrida*.
6 Eco, *Theory of Semiotics*, p. 7.

1. The Concept of Culture

1 Cited in Eagleton, *Literary Theory*, p. 20.
2 See Descartes, 'Meditations on First Philosophy' (L. 1642) in *Philosophical Writings*, pp. 59–124. For further discussion of Descartes, see my *Debating Derrida*, pp. 48–71.
3 Aristotle did, however, look upon *ethos* (in terms of human character and custom) as a condition to be struggled for. See his *Ethics*.
4 Hobbes, *Leviathan*, p. 185. For further discussion of Hobbes, see my *Postmodern Literary Theory*, pp. 46–9.
5 Callon and Latour, 'Unscrewing', p. 278. The embedded reference is to Hobbes, *Leviathan*, p. 227.
6 Vico, *New Science*, p. 33.
7 Ibid., p. 22.
8 Ibid., p. 21.
9 Ibid., p. 105.
10 Ibid., p. 244.
11 Ibid., p. 74.
12 Ibid., p. 112.

13   Ibid., p. 96.
14   Fisch, 'Introduction', p. xliv.
15   Hawkes, *Structuralism and Semiotics*, p. 15. My discussion of Vico draws
      liberally on Hawkes's discussion of his ideas; see esp. pp. 11–15.
16   Vico, *New Science*, p. 64.
17   Taine, *History of English Literature*, p. 7.
18   Stewart, *Does God Play Dice?*, p. 7.
19   See Herder, *Reflections*.
20   The question of cultural specificity is very much alive at present, in the
      gloomy context of the so-called globalization of contemporary culture. It
      could be, however, that the present (postmodern) condition has caused
      simply the forms of cultural specificity to change, but not to any totalizing
      effect; see Wark, *Virtual Geography* and *Celebrities*.

2. A Short History of Semiotics

1   Saussure, *Course*, p. 7.
2   Blonsky, 'Introduction', p. xxxvi.
3   Ibid.
4   In Saussure's terms, *langue* defines the inferred set of rules or principles of
      any semiotic system, although Saussure himself was interested exclusively
      in linguisitic systems. These principles are inferrable from the study of
      actual or concrete instances of semiotic events which, for Saussure,
      amounted to individual speech acts or what he termed *parole* forms of
      language. The *langue/parole* distinction is crucial to the whole Saussurean
      enterprise, but it is not without significant problems. 'What,' as Jonathan
      Culler neatly puts it, 'belongs to *langue* and what to *parole*?' ('Introduction',
      p. xxi). As we will see in later chapters, the particular problems associated
      with the *langue/parole* distinction bear on a more general form of what
      Derrida calls the speech/writing opposition, to which extent those pro-
      blems are always already beyond semiotics.
5   Saussure, *Course*, p. 9.
6   Blonsky, 'Introduction', p. xxvii.
7   For Saussure's 'syntagmatic and associative relations' we now use the
      terms 'syntagmatic and paradigmatic relations', referring to the means by
      which a sign or text is always a product of differences on the syntagmatic or
      horizontal plane of semiosis (e.g. 'c-a-t' is recognizable as the word 'cat'
      because 'c' is not 'a' which is not 't') and of selections made from within a
      set of possible associations on the paradigmatic or vertical plane (e.g. 'c-at'
      constitutes the word 'cat' because it is not 'b-at' or 'v-at', etc.). Hence a
      paradigm would be a menu, a wardrobe, a set of generic conventions, etc.,
      whereas the corresponding syntagm would be a meal, an outfit or an actual
      novel. Paradigmatic relations are only ever inferrable, then, but Saussure's

point is that they are nonetheless structurally vital to the production and exchange of sense; see Saussure, *Course*, pp. 122–7.

8   Blonsky, 'Introduction', p. 8.
9   Ibid.
10   Ibid., p. xlix.
11   Ibid., p. xviii.
12   Ibid., p. xl.
13   On the importance of nonlinguistic signs, whose recognition may be said to constitute the 'corporeal turn', see Ruthrof, *Body in Language* and *Semantics and the Body*. On Ruthrof's account, Saussure himself provides the correction to his own linguistic bias if his notion of reference is broadened to mean 'a relation between different semiotic systems by which we construct our world (within constraints)'; from this it would be possible to 'theorize meaning in a quasi-corporeal manner' (*Body in Language*, p. 156).
14   See Caputo, *Against Ethics*. To put the point generally, by way of an example: someone who took advantage of an opportunity to ruin someone else's marriage could not be condemned out of hand for having behaved 'unethically', as if there were some absolute standard or moral system of good and bad behaviours to which we could appeal. The most we could say (which is still a lot) is that the *decision* to take advantage of such an opportunity would not infer an *ethics* that many of us would care to endorse.
15   Blonsky, 'Introduction', p. l.
16   Derrida, *Gift of Death*, p. 36.

3. Total Eclipse of the Heart

1   See Plato, *Phaedrus*.
2   Derrida, *Dissemination*, p. 109.
3   For Saussure, a symbol designates a particular class of sign in which 'the rudiment of a natural bond' adheres between signifier and signified (*Course*, p. 68); hence the internal relations of the symbol are characterized by some degree of relative 'motivation', in contrast to the 'arbitrary' nature of the sign in general (ibid., pp. 131–4). The term 'index' derives from the founding American semiotician Charles Sanders Peirce and refers to the causal or contiguous relations of some signs (smoke as an index of fire); confusingly, the Peircean 'symbol' corresponds to the Saussurean (arbitrary) 'sign' (see Peirce, *Collected Papers*).
4   The term 'diegesis' has a long history in poetics, beginning with Plato in Book III of *The Republic* where it refers, in contrast to 'mimesis' (direct speech or representation), to the poetic style defined as the poet 'speaking in his own person' without any effort 'to persuade us that the speaker is anyone but himself' (p. 150). Significantly, Plato regards diegesis as the

'purest' form of narrative. Nowadays, for the discipline of narratology (heavily influenced by semiotics), diegesis refers to the story-form of any narrative text; that is, to the fictional or imagined world of events as inferred from the narrative (see Genette, *Narrative Discourse*, pp. 162–70).

5 Sofia, 'Exterminating fetuses', p. 57.
6 In Peirce's terminology an 'icon' defines a sign (such as a photograph) which is highly realistic or (in Saussure's terms) highly motivated.
7 Sofia, 'Exterminating fetuses', p. 49; emphasis added.

4. The Phake Fone

1 McLuhan, *Understanding Media*, pp. 286–8.
2 Ibid., p. 287.
3 Ibid.
4 Ibid.
5 Kant defines a noumenon as a thing or object in itself, 'an *object of a non-sensible intuition*' whose existence is independent of our perception of it; by comparison the term 'phenomena' designates 'certain objects as appearances or sensible existences . . ., thus distinguishing our mode of intuiting them from their own nature as things in themselves' (*Pure Reason*, p. 211). For further discussion on this topic, see Chapter 8.
6 McLuhan, *Understanding Media*, pp. 286–7.
7 Consider here the recent and perfectly postmodern event of Elvis's 1999 World Tour! A band comprising musicians famously associated with the King (James Burton, Ronnie Tutt, Glen D. Hardin *et al.*) plays live accompaniment to Elvis's recorded voice, his image video-projected on a giant screen in front of tens of thousands of adoring and no doubt unbemused fans in concert venues all around the world, the whole event involving legions of production, staging, lighting, publicity and other staff under the auspices of Graceland and Elvis Presley Enterprises, Inc.! For those who missed the show, consolation is at hand in the form of a CD recording of the tour's best moments, a tour that must be regarded simply as Elvis's latest and proving that the greatest Elvis impersonator of them all is Elvis himself. Even the finality of death, it seems, is not quite what it used to be; see also Chapter 6.
8 Brunette and Wills, *Screen/Play*, p. 183.
9 Gumpert, *Talking Tombstones*, p. 129.
10 Moyal, 'Women and technology', p. 57.
11 Ibid.
12 Ibid.
13 White, 'Immigrants', p. 61.
14 Ibid.
15 Evans, *Telecommunications*, p. 3.

16 Ibid.
17 Ibid., p. 189.
18 Ibid.
19 Moyal, 'Women and technology', p. 58.
20 Gumpert, *Talking Tombstones*, p. 130.
21 Ibid., p. 131.
22 Ibid., p. 132.
23 See Derrida, *Post Card*.
24 Gumpert, *Talking Tombstones*, p. 132.

5. Situating Technologies

1 Derrida, *Post Card*, p. 68.
2 The postal metaphor alludes to the necessary errancy and uncontrollability of meaning, on the model of the postal system which can never guarantee the safe arrival of a letter at its intended address. The postal system is subject to a range of mailing mishaps (dead and misaddressed letters, false signatories, etc.), which in fact constitute the condition by which any letter may be said to have arrived as intended. Safe arrival, then, is actually a special kind of accident, contrary to what we might regard as our experience of the postal system in good working order. By analogy (and here we are moving well beyond semiotics, without quite leaving semiotics behind), there is no 'destined' relation between any signifier and the signified associated with it. But while Saussure's concept of the arbitrariness of the sign allows for this principle of adestination, semiotics (like a good postal system) has tended to suppress or ignore this stubborn condition, preferring to treat signs as if they operated within only a restricted economy of perfect sense and as if they remained forever unexposed to any aleatory effects of meaning. See Derrida, *Post Card*; Norris, *Derrida*, pp. 185–93 and Lucy, *Debating Derrida*, pp. 89–92. See also Chapters 7, 8 and 9.
3 'Logocentrism' refers to the metaphysical tradition (epitomized by philosophy but common to all forms of Western thinking) by which truth and meaning are understood as foundational, even noumenal. To this extent truth is understood as being always present to itself; hence semiotics represents a potentially radical critique of logocentrism in so far as, for Saussure, truth is not positive in its own right but is an effect only of a system of differences, a system without positive terms. By the same token Saussure's reliance on the seemingly self-evidential nature of the difference between this signified and that one, or between this speech act and that rule-governing principle, is stubbornly logocentric in its faith in the self-presence and self-constitution of individual entities. Even more revealing of semiotics' deep-seated logocentrism is its confidence in 'knowing' the

signified by means of access through the signifier, as though signifiers were but the outer 'intelligible' (or phenomenal) layers of deep 'sensible' (or noumenal) truths. In our refusal of that confidence here we're attempting to go beyond semiotics and so (in a sense) beyond logocentrism, although this statement needs to be set against Derrida's important reminder that there could be 'no sense in doing without the concepts of metaphysics in order to shake metaphysics' (*Writing and Difference*, p. 280); see also Derrida, *Grammatology* and Culler, *Deconstruction*, pp. 92–111.

4   For the German philosopher and social theorist Walter Benjamin, the technological reproduction of artworks in the twentieth century actually helps to produce and certainly to maintain the 'aura' surrounding original works as well as the concept of the 'original' itself; see 'Work of art'.

5   Berger, *Ways of Seeing*, p. 7.

6   Brecht, 'Radio', pp. 51–2.

7   Ibid., p. 52.

8   See Heidegger, 'Origin'.

9   Heidegger, *History of the Concept*, p. 232; emphasis added.

10 See Derrida, *Of Spirit*.

11 See Derrida, *Grammatology*.

12 'Phase space' represents the full extent of possibilities for any dynamical system; hence the state of such a system at any suspended moment would appear as a point in phase space. A 'strange attractor' is a force or point within a dynamical system that 'attracts' eccentric or 'chaotic' behaviour. See Gleick, *Chaos*.

13 Derrida, *Writing and Difference*, p. 293.

14 Ibid. The term 'differance' is an Anglicized variant of the French neologism *différance* deriving from the verb *différer* meaning to differ and to defer. Hence differance is both 'a structure and a movement', as Derrida puts it (*Positions*, p. 27), which is actively productive of differences and whose nature (as a structure and a movement) is inconceivable 'on the basis of the opposition presence/absence' (ibid.). Derrida gives the example of *langue* and *parole*, noticing that any rigorous distinction between these terms raises the question of which comes first, the rule-generating system or the speech acts which, in order to be intelligible, require the system to instantiate itself. 'Therefore, one has to admit, before any dissociation of language and speech [*langue* and *parole*], code and message, etc. (and everything that goes along with such a dissociation), a systematic production of differences, the *production* of a system of differences – a *différance* – within whose effects one eventually, by abstraction and according to determined motivations, will be able to demarcate a linguistics of language and a linguistics of speech, etc.' (ibid., p. 28). See Derrida, 'Différance'; Caputo, *Nutshell*, pp. 96–105 and Lucy, *Debating Derrida*, pp. 59–71.

15 This argument is developed at length in my *Postmodern Literary Theory*, see esp. pp. 141–62.

16 Sofia, 'Exterminating fetuses', p. 57.
17 Ibid., p. 49.
18 Ibid., p. 57.
19 Ibid.
20 See esp. Nietzsche, *Genealogy*.
21 Derrida, *Other Heading*, p. 103.
22 Cited in Frith, 'Pleasures of the hearth', p. 103.
23 See Plato, *Phaedrus* and Derrida, *Dissemination*.
24 Tofts, *Parallax*, p. 9.
25 See Derrida, *Writing and Difference*, pp. 278–93 and Lucy, *Postmodern Literary Theory*, pp. 97–103; on the inventiveness of scientists and mathematicians, see Gleick, *Chaos* and Stewart, *Does God Play Dice?*

6. The Sound of a Dream

1 See Derrida, *Dissemination*, pp. 61–171 and Frow, *Time and Commodity Culture*, pp. 218–46.

7. Catholic English

1 These lines are taken from a song called 'Kodachrome' by Paul Simon (whom I vow never to cite again!).
2 James, *Turn of the Screw*.
3 There are at least three exceptions: Felman, 'Turning the screw'; Brooke-Rose, *Rhetoric of the Unreal* and Anderson, '"Fury of intention"'.
4 Robbins, *Another Roadside Attraction*, p. 36.
5 But see the final chapter, 'Is "pi" really in the sky?,' in Barrow, *Theories of Everything*. As the chapter title indicates, the question of whether pi is real or simply functional is a source of ongoing debate within mathematics: 'The choice requires us to know whether the laws of physics constrain the ultimate capability of abstract computation. Do they limit its speed and scope? Or do the rules governing the process of computation control what laws of Nature are possible?' (p. 204). In philosophical terms, the question would be whether statements about pi are analytic (that is, logically necessary but not otherwise, or not always, true) or synthetic (always already true); but there is no such debate within mathematics over the purely analytic status of statements concerning $\sqrt{1}$, which might therefore be a better example at this point.
6 See Gleick, *Chaos*.
7 Milton wrote the pastoral elegy 'Lycidas' in honour of Edward King, a young Cambridge graduate who drowned in the Irish Sea in August 1637.

8   Nietzsche, *Anti-Christ*, p. 157.
9   The canon (Greek *kanon*, 'measuring rod, rule') refers strictly to those books of the Bible which conform to the rule of holy or divinely inspired authorship, so that the canon is both the rule or set of rules of scholarship by which the legitimacy of the biblical text is determined, and the biblical text itself – it is both literature *and* criticism. By contrast, the apocrypha are those books considered not to measure up to canonical standard, being of spurious (unruly or 'illegitimate') authorship or indeterminate origin; hence to argue their inclusion in the Bible (the canon) would be an act of bad faith or bad criticism (and therefore uncanonical) since they are of themselves uncanonical to begin with. The question then is whether the rule of the Bible's determination is what determines the Bible, or whether the Bible determines it (the rule). In any case, it is clear that the canon is used to determine authorship *and* to describe the works so determined, though it remains undecidable as to whether the Word of God is a product of human justification or divine inscription. As a rule, does the canon constrain the conditions of the canon as text? Or is the rule (the canon) a divine (or natural) law, fully capable of realizing the canon as text? More to the point, how could the canonites decide? See also Leavis, *Great Tradition*.
10 Freadman and Miller, 'Three views', p. 24.
11 Foucault, *Archaeology*, p. 5. The embedded reference is to Althusser, *For Marx*, p. 168.
12 See the famous debate in *Critical Inquiry* between M. H. Abrams and J. Hillis Miller over the nature of text–commentary relations: Abrams, 'Deconstructive angel' and Miller, 'Critic as host'.
13 Abrams, 'Deconstructive angel', p. 433.
14 Ibid., p. 432.
15 Cited in Shakespeare, *Arden Shakespeare*, p. 205.
16 Shakespeare, *Stratford Shakespeare*.
17 In 'A Note on the Text' to a collection of Melville's stories, Harold Beaver quotes from a letter dated 24 March 1856 from the author to his publishers, Dix and Edwards, freely lending permission to edit his work according to their better judgement: 'There seems to be a surprising profusion of commas in these proofs,' Melville wrote. 'I have struck them out pretty much; but I hope that some one who understands punctuation better than I do, will give the final hand to it' (cited in Melville, *Billy Budd*, p. 51). 'Occasionally,' Beaver continues, 'with this for cue, I have deleted a comma, or even added one, as sense demanded' (ibid.).
18 Perkins was not alone in this; Fitzgerald's grammatical eccentricities were corrected also by the editors of the magazines that bought his stories. See Bruccoli, 'Introduction'.
19 Chambers, *Room for Maneuver*, p. xii.
20 Kamuf, *Division of Literature*, p. 17.

8. Derivations

1   See Heidegger, *Introduction to Metaphysics* and *Basic Questions*.
2   Empson, *Ambiguity*, pp. xiii–xv.
3   Cited in Empson, *Ambiguity*, p. xii.
4   Lucy and McHoul, 'Lowry's envois', p. 6.
5   See Foucault, 'What is an author?'
6   Derrida, *Grammatology*, p. 91.
7   Besides, others have come before me. For an account of the 'error messages' in Spivak's translation of the *Grammatology*, see McHoul and Wills, *Writing Pynchon*, esp. chap. 4, 'Telegrammatology'.

9. Gilligans Wake

1   Barthes, 'Soap powders', p. 36.
2   Ulmer, *Teletheory*, p. 63 and passim.
3   Ulmer, *Teletheory*, p. 150; Derrida, *Positions*, p. 43.
4   McHoul and Wills, *Writing Pynchon*, p. 108.
5   Ibid., p. 109.
6   Ibid., p. 112.
7   Ibid.
8   See Derrida, *Grammatology*.
9   Barthes, 'From work to text,' pp. 159–60.
10   Ulmer, *Teletheory*, p. 21.
11   See Hartley, *Tele-ology* and *Uses of Television*.
12   McHoul and Wills, *Writing Pynchon*, p. 112.
13   Ibid., p. 113.
14   Barthes, *Camera Lucida*, p. 27.
15   McHoul and Wills, *Writing Pynchon*, p. 216.
16   Ibid., pp. 216–17.
17   Derrida, *Grammatology*, pp. 22, 23.
18   Ulmer, *Teletheory*, pp. 15–16.
19   Lévi-Strauss, *Structural Anthropology*, p. 214.
20   Ibid., p. 213.
21   Ibid., p. 217.
22   Ibid., p. 218.
23   McHoul and Wills, *Writing Pynchon*, p. 114.
24   Ibid., p. 115.
25   Ulmer, *Teletheory*, p. 43.
26   Ibid., p. vii.
27   Hartley, *Tele-ology*, p. 11.
28   Ulmer, *Teletheory*, pp. 73–9.

29   Ibid., p. 209.
30   Aarseth, *Cybertext*, p. 172.

Coda: Interzones

1   Maddox, 'Snake-Eyes', p. 27. Further references cited in the text.
2   See Porush, *Soft Machine*, pp. 56–8.
3   See Weiner, *Human Use*. The uncertainty principle, attributed to the quantum physicist Werner Heisenberg, encapsulates (among other ideas) the notion that the outcome of any measurement is affected by the act of measurement itself.
4   Tofts and McKeich, *Memory Trade*, p. 30.
5   Wark, *Celebrities*, p. 65.
6   Haraway, 'Manifesto for cyborgs', p. 83.
7   Wark, *Celebrities*, p. 333.
8   Derrida, *Post Card*, p. 311.
9   But see Caputo, *Prayers and Tears*.
10   Saussure, *Course*, p. 20.
11   This is not to say that there is nothing whatsoever outside of language (see for example Ruthrof, *Body in Language* and Merleau-Ponty, *Phenomenology*), but rather that there's no outside to textuality or semiosis; in a word, there is no outside-context.
12   Caputo, *Nutshell*, p. 100.
13   See Derrida, *Specters*.
14   See Ott, *Heidegger*. This is not in any way intended to redeem Heidegger by inferring that he was somehow an enemy of national socialism and that's the reason the ban was imposed. Far from it, but that is not the point for now.
15   See Deleuze and Guattari, *A Thousand Plateaus*. For a discussion of some of their ideas, including nomadology, see my *Postmodern Literary Theory*, pp. 184–226.
16   See Hartley, *Popular Reality* and *Uses of Television*; Lucy, *Philosophy and Cultural Studies* and Wark, *Thinking Media*.
17   See Briggs, 'Derrida and the question of discipline'.

Bibliography

Aarseth, Espen J. *Cybertext: Perspectives on Ergodic Literature*. Baltimore: Johns Hopkins University Press, 1997.

Abrams, M. H. 'The limits of pluralism, II: The deconstructive angel', *Critical Inquiry* 3 (Spring 1977), pp. 425–38.

Althusser, Louis. *For Marx*, trans. Ben Brewster. New York: Pantheon, 1969.

Anderson, Don. '"A fury of intention": the scandal of Henry James' *The Turn of the Screw*', *Sydney Studies in English* 15 (1989–90), pp. 140–52.

Aristotle, *The Ethics of Aristotle: The Nichomachean Ethics*, trans. J. A. K. Thomson and Hugh Tredennick. Harmondsworth: Penguin, 1976.

Barrow, John D. *Theories of Everything: The Quest for Ultimate Explanation*. Oxford: Clarendon, 1991.

Barthes, Roland. 'From work to text', in *Image Music Text*, trans. Stephen Heath. London: Flamingo, 1982, pp. 155–64.

Barthes, Roland. 'Soap powders and detergents', in *Mythologies*, trans. Annette Lavers. London: Paladin, 1983, pp. 36–8.

Barthes, Roland. *Camera Lucida: Reflections on Photography*, trans. Richard Howard. London: Flamingo, 1984.

Benjamin, Walter. 'The work of art in the age of mechanical reproduction', in *Illuminations*, trans. Harry Zohn. London: Fontana, 1973, pp. 219–53.

Berger, John. *Ways of Seeing*. Harmondsworth: Penguin, 1977.

Blonsky, Marshall. 'Introduction: the agony of semiotics: reassessing the discipline', in Marshall Blonsky (ed.), *On Signs: A Semiotics Reader*. Oxford: Blackwell, 1985, pp. xiii–li.

Brecht, Bertold. 'The radio as an apparatus of communication', in *Brecht on Theatre: The Development of an Aesthetic*, trans. and ed. John Willett. London: Methuen, 1987, pp. 51–3.

Briggs, Robert. 'Derrida and the Question of Discipline'. Unpublished PhD thesis. Murdoch University, 2000.

Brooke-Rose, Christine. *A Rhetoric of the Unreal: Studies in Narrative & Structure, Especially of the Fantastic*. Cambridge: Cambridge University Press, 1981.

Bruccoli, Matthew J. 'Introduction' to F. Scott Fitzgerald, *The Great Gatsby*. London: Abacus, 1992, pp. vii–xx.

Brunette, Peter and David Wills. *Screen/Play: Derrida and Film Theory*. Princeton: Princeton University Press, 1988.

Callon, Michael and Bruno Latour. 'Unscrewing the big Leviathan: how actors macro-structure reality and how sociologists help them to do so', in K. Knorr-Cetina and A. Cicourel (eds), *Advances in Social Theory and Methodology*. London: Routledge and Kegan Paul, 1981, pp. 277–303.

Caputo, John D. *Against Ethics: Contributions to a Poetics of Obligation with Constant Reference to Deconstruction*. Bloomington: Indiana University Press, 1993.

Caputo, John D. *Deconstruction in a Nutshell: A Conversation with Jacques Derrida*. New York: Fordham University Press, 1997.

Caputo, John D. *The Prayers and Tears of Jacques Derrida: Religion Without Religion*. Bloomington: Indiana University Press, 1997.

Chambers, Ross. *Room for Maneuver: Reading (the) Oppositional (in) Narrative*. Chicago: University of Chicago Press, 1991.

Culler, Jonathan. 'Introduction', in Ferdinand de Saussure, *Course in General Linguistics*, trans. Wade Baskin. London: Fontana, 1974, pp. xi–xxv.

Culler, Jonathan. *On Deconstruction: Theory and Criticism after Structuralism*. London: Routledge and Kegan Paul, 1985.

Danesi, Marcel. *Messages and Meanings: An Introduction to Semiotics*. Toronto: Canadian Scholars' Press, 1993.

Deleuze, Gilles and Félix Guattari. *A Thousand Plateaus: Capitalism and Schizophrenia*, trans. Brian Massumi. Minneapolis: University of Minnesota Press, 1987.

Derrida, Jacques. 'Différance', in *Margins of Philosophy*, trans. Alan Bass. Chicago: Chicago University Press, 1982, pp. 3–27.

Derrida, Jacques. *Dissemination*, trans. Barbara Johnson. Chicago: Chicago University Press, 1981.

Derrida, Jacques. *Of Grammatology*, trans. Gayatri Chakravorty Spivak. Baltimore: Johns Hopkins University Press, 1976.

Derrida, Jacques. *Of Spirit: Heidegger and the Question*, trans. Geoffrey Bennington and Rachel Bowlby. Chicago: University of Chicago Press, 1989.

Derrida, Jacques. *Positions*, trans. Alan Bass. Chicago: University of Chicago Press, 1981.

Derrida, Jacques. *Specters of Marx: The State of the Debt, the Work of Mourning, and the New International*, trans. Peggy Kamuf. New York: Routledge, 1994.

Derrida, Jacques. *The Gift of Death*, trans. David Wills. Chicago: University of Chicago Press, 1995.

Derrida, Jacques. *The Other Heading: Reflections on Today's Europe*, trans. Pascale-Anne Brault and Michael B. Naas. Bloomington: Indiana University Press, 1992.

Derrida, Jacques. *The Post Card: From Socrates to Freud and Beyond*, trans. Alan Bass. Chicago: University of Chicago Press, 1987.

Derrida, Jacques. *Writing and Difference*, trans. Alan Bass. London: Routledge and Kegan Paul, 1978.

Descartes, René *Philosophical Writings*, trans. Elizabeth Anscombe and Peter Thomas Geach. London: Thomas Nelson, 1970.

Eagleton, Terry. *Literary Theory: An Introduction*. 2nd edn. Oxford: Blackwell, 1996.

Eco, Umberto. *A Theory of Semiotics*. Bloomington: Indiana University Press, 1976.

Empson, William. *Seven Types of Ambiguity*. London: Hogarth, 1947.

Evans, Gareth. *Australian Telecommunications Services: A New Framework*. Canberra: Australian Government Printing Services, 25 May 1988.

Felman, Shoshana. 'Turning the screw of interpretation', *Yale French Studies: Literature and Psychoanalysis. The Question of Reading: Otherwise* 55/56 (1977), pp. 94–207.

Fisch, Max Harold. 'Introduction' to Giambattista Vico, *The New Science of Giambattista*

Vico, 3rd edn (*1744*), trans. Thomas Goddard Bergin and Max Harold Fisch. Ithaca, NY: Cornell University Press, 1984, pp. xix–xlv.

Foucault, Michel. 'What is an author?', trans. José V. Harari, in *The Foucault Reader,* ed. Paul Rabinow. Harmondsworth: Penguin, 1986, pp. 101–20.

Foucault, Michel. *The Archaeology of Knowledge,* trans. A. M. Sheridan Smith. London: Routledge, 1972.

Freadman, Richard and S. R. Miller. 'Three views of literary theory' in *Literary Theory in Australia,* D. Freundlieb (ed.), *Poetics* 17, 1–2 (April 1988), pp. 9–24.

Frith, Simon. 'The pleasures of the hearth: the making of BBC light entertainment', in Formations Collective (ed.), *Formations of Pleasure.* London: Routledge and Kegan Paul, 1983, pp. 101–23.

Frow, John. *Time and Commodity Culture: Essays in Cultural Theory and Postmodernity.* Oxford: Clarendon Press, 1997.

Genette, Gérard. *Narrative Discourse,* trans. Jane E. Lewin. Oxford: Blackwell, 1980.

Gleick, James. *Chaos: Making a New Science.* London: Cardinal, 1988.

Gottdiener, M. *Postmodern Semiotics: Material Culture and the Forms of Postmodern Life.* Oxford: Blackwell, 1995.

Gumpert, Gary. *Talking Tombstones and Other Tales of the Media.* New York: Oxford University Press, 1987.

Haraway, Donna. 'A manifesto for cyborgs: science, technology, and socialist feminism in the 1980s', *Socialist Review* 15, 2 (1985), pp. 65–108.

Hartley, John. *Popular Reality: Journalism, Modernity, Popular Culture.* London: Arnold, 1996.

Hartley, John. *Tele-ology: Studies in Television.* London: Routledge, 1992.

Hartley, John. *Uses of Television.* London: Routledge, 1999.

Hawkes, Terence. *Structuralism and Semiotics.* London: Routledge, 1983.

Heidegger, Martin. 'Origin of the work of art', trans. Albert Hofstadter, in *Poetry, Language, Thought.* New York: Harper and Row, 1971, pp. 17–87.

Heidegger, Martin. *An Introduction to Metaphysics,* trans. Ralph Manheim. New Haven: Yale University Press, 1987.

Heidegger, Martin. *Basic Questions of Philosophy: Selected 'Problems' of 'Logic',* trans. R. Rojcewicz and A. Schuwer. Bloomington and Indianapolis: Indiana University Press, 1984.

Heidegger, Martin. *History of the Concept of Time,* trans. Theodore Kisiel. Bloomington: Indiana University Press, 1985.

Herder, Johann Gottfried. *Reflections on the Philosophy of the History of Mankind,* trans. F. E. Manuel. Chicago: University of Chicago Press, 1969.

Hobbes, Thomas. *Leviathan,* ed. C. B. Macpherson. Harmondsworth: Penguin, 1968.

James, Henry. *The Aspern Papers and The Turn of the Screw,* ed. Anthony Curtis. Harmondsworth: Penguin, 1984.

Kamuf, Peggy. *The Division of Literature, or, The University in Deconstruction.* Chicago: University of Chicago Press, 1997.

Kant, Immanuel. *Critique of Pure Reason,* trans. Vasilis Politis (based on translation by J. M. D. Meikljohn). London: J. M. Dent, 1993.

Leavis, F. R. *The Great Tradition: George Eliot, Henry James, Joseph Conrad.* Harmondsworth: Penguin, 1962.

Lévi-Strauss, Claude. *Structural Anthropology 1,* trans. Claire Jacobson and Brooke Grundfest Schoepf. Harmondsworth: Penguin, 1968.

Lucy, Niall and Alec McHoul. 'Lowry's envois', *Sub-Stance* 70 (1993), pp. 3–24.

Lucy, Niall (ed.). *Philosophy and Cultural Studies: Continuum* 12, 2 (July 1998).

Lucy, Niall. 'Introduction (on the way to genre)', in Niall Lucy (ed.), *Postmodern Literary Theory: An Anthology*. Oxford: Blackwell, 2000, pp. 1–39.

Lucy, Niall. *Debating Derrida*. Carlton: Melbourne University Press, 1995.

Lucy, Niall. *Postmodern Literary Theory: An Introduction*. Oxford: Blackwell, 1997.

McHoul, Alec and David Wills. *Writing Pynchon: Strategies in Fictional Analysis*. London: Macmillan, 1990.

McLuhan, Marshall. *Understanding Media: The Extensions of Man*. London: Sphere, 1967.

Melville, Herman. *Billy Budd, Sailor and Other Stories*, ed. Harold Beaver. Harmondsworth: Penguin, 1983.

Merleau-Ponty, Maurice. *Phenomenology of Perception*, trans. Colin Smith. London: Routledge and Kegan Paul, 1962.

Mickler, Steve. *The Myth of Privilege: Aboriginal Status, Media Visions, Public Ideas*. Fremantle: Fremantle Arts Centre Press, 1998.

Miller, J. Hillis. 'The limits of pluralism, III: The critic as host', *Critical Inquiry* 3 (Spring 1977), pp. 439–47.

Moyal, Anne. 'Women and technology: a case study of the telephone in Australia', *Media Information Australia* 54 (1989), pp. 57–60.

Nietzsche, Friedrich. *On the Genealogy of Morality*, trans. Carol Diethe. New York: Cambridge, 1994.

Nietzsche, Friedrich. *Twilight of the Idols and The Anti-Christ*, trans. R. J. Hollingdale. Harmondsworth: Penguin, 1977.

Norris, Christopher. *Derrida*. London: Fontana, 1987.

Ott, Hugo. *Martin Heidegger: A Political Life*, trans. Allan Blunden. London: Fontana, 1994.

Peirce, Charles S. *Collected Papers of Charles Sanders Peirce*, ed. Charles Hartshome and Paul Weiss. Cambridge, MA: Harvard University Press, 1965.

Plato. *Phaedrus*, trans. W. C. Helmbold and W. G. Rabinowitz. Indianapolis: Bobbs-Merrill Educational Publishing, 1956.

Porush, David. *The Soft Machine: Cybernetic Fiction*. London: Methuen, 1985.

Robbins, Tom. *Another Roadside Attraction*. New York: Ballantine, 1980.

Ruthrof, Horst. *Semantics and the Body: Meaning from Frege to the Postmodern*. Toronto: University of Toronto Press, 1997.

Ruthrof, Horst. *The Body in Language*. London: Cassell, 2000.

Saussure, Ferdinand de. *Course in General Linguistics*, trans. Wade Baskin. London: Fontana, 1974.

Shakespeare William. *The Illustrated Stratford Shakespeare*. London: Chancellor Press, 1989.

Shakespeare, William. *The Arden Shakespeare: King Lear*, ed. Kenneth Muir. London: Methuen, 1975.

Sofia, Zoë. 'Exterminating fetuses: abortion, disarmament, and the sexo-semiotics of extraterrestrialism', *Diacritics* 14, 2 (1984), pp. 47–59.

Stewart, Ian. *Does God Play Dice?* Harmondsworth: Penguin, 1990.

Taine, Hippolyte. *History of English Literature*, Vol. 1, trans. H. Van Laun. London: Chatto and Windus, 1907.

Tofts, Darren and Murray McKeich. *Memory Trade: A Prehistory of Cyberculture*. Sydney: Interface, 1997.

Tofts, Darren. *Parallax: Essays on Art, Culture and Technology*. Sydney: Interface, 1999.

Ulmer, Gregory. *Teletheory: Grammatology in the Age of Video*. London: Routledge, 1989.

Vico, Giambattista. *The New Science of Giambattista Vico*, 3rd edn (1744), trans. Thomas Goddard Bergin and Max Harold Fisch. Ithaca, NY: Cornell University Press, 1984.

Wark, McKenzie (ed.). *Thinking Media*. Sydney: Pluto, 2000.

Wark, McKenzie. *Celebrities, Culture and Cyberspace: The Light on the Hill in a Postmodern World*. Sydney: Pluto, 1999.

Wark, McKenzie. *Virtual Geography: Living with Global Media Events*. Bloomington: Indiana University Press, 1994.

Weiner, Norbert. *The Human Use of Human Beings: Cybernetics and Society*. London: Sphere, 1968.

White, Peter. 'Immigrants and the telephone in Australia', *Media Information Australia* 54 (1989), pp. 61–6.

Index